Thin Film Transistor Circuits and S

T0332881

Providing a reliable and consolidated treatment of the principles behind large-area electronics, this book contains a comprehensive review of the design challenges associated with building circuits and systems from thin film transistors.

The authors describe the architecture, fabrication, and design considerations for the principal types of TFT, and their numerous applications. The practicalities of device non-ideality are also addressed, as are the specific design considerations necessitated by instabilities and non-uniformities in existing fabrication technologies.

Containing device-circuit information, discussion of electronic solutions that compensate for material deficiencies, and design methodologies applicable to a wide variety of organic and inorganic disordered materials, this is an essential reference for all researchers and circuit and device engineers working on large-area electronics.

Reza Chaji is the President and Chief Technology Officer at Ignis Innovation Inc., managing programs for the development of low-cost, high-yield, and low-power AMOLED displays. He is an Adjunct Professor at the University of Waterloo, Canada, and has been awarded the CMC Douglas R. Colton Medal for Research Excellence.

Arokia Nathan is the Chair of Photonic Systems and Displays at the University of Cambridge, former Sumitomo Chair of Nanotechnology at the London Centre for Nanotechnology, University College London, and the Canada Research Chair, University of Waterloo. The founder or co-founder of several companies, he is also a Fellow of the IEEE and the IET, and an IEEE/EDS Distinguished Lecturer.

"The book is an absolute must for everyone seriously interested in pixel circuits for active matrix organic light-emitting displays and flat panel imagers."

Norbert Fruehauf, University of Stuttgart

"Various TFT materials and devices have been developed for addressing liquid crystal displays and organic light-emitting diode displays. Lately, high mobility oxide semiconductors are emerging, which promises higher resolution and larger aperture ratio for improving optical efficiency. There is an urgent need for such a book to give a systematic approach to basic material properties, advanced circuit designs, and integrated operation systems for this $100B display industry. The authors are respected experts in the field. I wish to have this book on my desk soon."

Shin-Tson Wu, University of Central Florida

Thin Film Transistor Circuits and Systems

Reza Chaji
Ignis Innovation Inc.

Arokia Nathan
University of Cambridge

CAMBRIDGE
UNIVERSITY PRESS

Shaftesbury Road, Cambridge CB2 8EA, United Kingdom

One Liberty Plaza, 20th Floor, New York, NY 10006, USA

477 Williamstown Road, Port Melbourne, VIC 3207, Australia

314–321, 3rd Floor, Plot 3, Splendor Forum, Jasola District Centre, New Delhi – 110025, India

103 Penang Road, #05–06/07, Visioncrest Commercial, Singapore 238467

Cambridge University Press is part of Cambridge University Press & Assessment, a department of the University of Cambridge.

We share the University's mission to contribute to society through the pursuit of education, learning and research at the highest international levels of excellence.

www.cambridge.org
Information on this title: www.cambridge.org/9781107012332

First published 2013

A catalogue record for this publication is available from the British Library

Library of Congress Cataloging-in-Publication data
Chaji, Reza.
Thin film transistor circuits and systems / Reza Chaji, Arokia Nathan.
    pages   cm
ISBN 978-1-107-01233-2 (Hardback)
1. Thin film transistors.   2. Transistor circuits.   I. Nathan, Arokia, 1957–   II. Title.
TK7871.96.T45C43 2013
621.3815–dc23   2012046860

ISBN   978-1-107-01233-2   Hardback

Contents

Preface

Advances in thin film materials and process technologies continue to fuel new areas of application in large area electronics. However, this does not come without new issues related to device-circuit stability and uniformity over large areas, placing an even greater need for new driving algorithms, biasing techniques, and fully compensated circuit architectures. Indeed, each application is unique and mandates specific circuit and system design techniques to deal with materials and process deficiencies. As this branch of electronics continues to evolve, the need for a consolidated source of design methodologies has become even more compelling. Unlike classical circuit design approaches where trends are toward transistor scaling and high integration densities, the move in large-area electronics is toward increased functionality, in which device sizes are not a serious limitation. This book is written to address these challenges and provide system-level solutions to electronically compensate for these deficiencies.

Although the circuit and system implementation examples given are based primarily on amorphous silicon technology, the design techniques and solutions described are unique, and applicable to a wide variety of disordered materials, ranging from polysilicon and metal oxides to organic families. These are complemented by real-world examples related to active-matrix organic light emitting diode displays, bio-array sensors, and flat-panel biomedical imagers. We address mixed-phase thin film and crystalline silicon electronics and, in particular, the design and interface techniques for high and low voltage circuits in the respective design spaces. The content is concise but diverse, starting with an introduction to displays, flat panel imagers, and associated backplane technologies, followed by design specifications and considerations addressing compensation and driving schemes. Here we introduce hybrid voltage-current programming, enhanced-settling current programming, and charge-based driving schemes for high-resolution pixelated architectures.

Apart from designers of imaging and display systems and the engineering community at large, this book will benefit material scientists, physicists, and chemists working on new materials for thin film transistors and sensors. It can serve as a text or reference for senior undergraduate and graduate courses in electrical engineering, physics, chemistry, or materials science. Much of the material in the book can be presented in about 30 hours of lecture time.

This book would not have been possible without the support of the Giga-to-Nano Labs at the University of Waterloo; Ignis Innovation Inc. in Waterloo; the London Centre for Nanotechnology, University College London; and the Centre for Large Area Electronics, University of Cambridge. We acknowledge the financial support provided by the Natural Sciences and Engineering Research Council, Canada, the Communications and Information Technology Ontario, Canada, and the Royal Society Wolfson Merit Award, UK.

Reza Chaji and Arokia Nathan
Waterloo and Cambridge, 2013

1 Introduction

We are witnessing a new generation of applications of thin film transistors (TFTs) for flat-panel imaging [1, 2, 3] and displays [4, 5, 6]. Unlike the active matrix liquid crystal display (AMLCD) where the TFT acts as a simple switch [7], new application areas are emerging, placing demands on the TFT to provide analog functions including managing instability arising from material disorder [3, 6].

In the following sections, we briefly describe the application platforms we have considered in this book, namely flat-panel displays and imaging, along with a summary of performance characteristics of the key TFT technologies used, or being considered, by the large-area electronics industry. While the circuit architectures reported here use examples based on amorphous silicon technology, they are easily adaptable to a broad range of materials families and applications with different specifications.

1.1 Organic light emitting diode displays

OLEDs have demonstrated promising features to provide high-resolution, potentially low-cost, and wide-viewing angle displays. More importantly, OLEDs require a small current to emit light along with a very low operating voltage (3–10 V), leading to very power efficient light emitting devices [4–6].

OLEDs are fabricated either by organic (small molecule) or polymeric (long molecule) materials. Small molecule OLEDs are produced by an evaporation technique in a high vacuum environment [8],

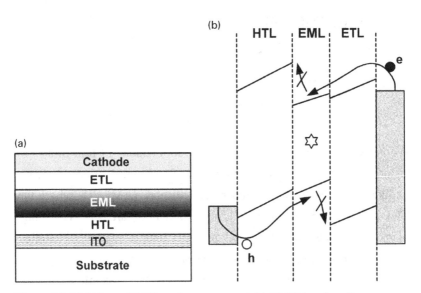

Figure 1.1 Multi-layer OLED stack structure and (b) OLED banding diagram adapted from [8, 10].

whereas, polymeric OLEDs are fabricated by spin-coating or inkjet printing [9]. However, the efficiency of small molecule OLEDs is much higher than that of polymeric OLEDs.

To increase the efficiency of the OLED, an engineered band structure is adopted [8]. A typical multi-layer OLED and its corresponding banding diagram are illustrated in Figure 1.1. The indium tin oxide (ITO) layer is the anode contact. The hole-transport layer (HTL), a p-doped layer, provides holes for the emission layer (EML), and also prevents electrons from traveling to the anode because of the band offset with the adjacent layers. For the cathode, the electron transport layer, an n-doped layer, provides electrons for the EML, and prevents the holes from traveling to the cathode. Then, the electrons and holes are recombined in the EML layer, resulting in the generation of photons [8, 10].

The luminance of OLEDs is linearly proportional to their current at low-to-mid current densities, and saturates at higher current densities.

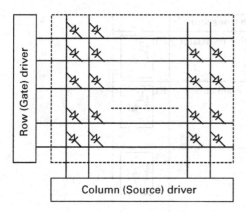

Figure 1.2 Passive matrix OLED display structure adapted from [16].

However, the voltage of OLEDs increases over time due to crystalliza-tion, chemical reaction at the boundaries, changes in the charge profile of the layer, and oxidation due to the existence of oxygen and moisture [11, 13]. Consequently, most of the proposed driving schemes are designed to provide a constant current for OLEDs.

OLEDs offer great promise in either passive or active formats. Figure 1.2 portrays passive matrix OLED (PMOLED) architecture. By applying a voltage across the appropriate row and column contacts, a specific pixel is addressed. Thereby, a current flows through the organic layers at the intersection of these contacts to light up the pixel. In this architecture, the luminance during the programming is averaged for the entire frame rate. Thus, the pixel should be programmed for $N \times L$ where N is the number of rows and L is the desired luminance for a frame [15, 16]. Thus, the OLED current density increases signifi-cantly, especially for higher resolution displays [5, 17]. Since the OLED efficiency drops at high current densities [18], to increase the display resolution, the current increases by a power law instead of linearly. Thus, the power consumption increases and the OLED ages faster. As a result, the actual applications of PMOLED displays are limited to small displays that have a low resolution [5].

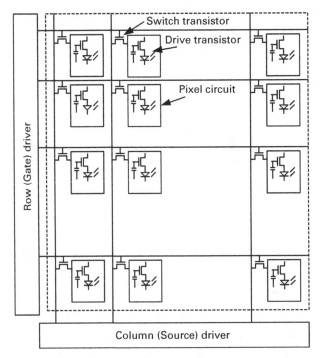

Figure 1.3 Active matrix OLED (AMOLED) display structure.

To increase the resolution and area of the displays, active matrix addressing is selected [5]. A simplified active matrix OLED (AMOLED) display structure is illustrated in Figure 1.3, where the pixel current is controlled by a drive transistor. During the programming cycle, the switch TFT is ON, and the pixel data is stored in the storage capacitor. During the driving cycle, a current, related to the stored data voltage, is provided to the OLED. Since the pixel current is smaller in the AMOLED displays, they have longer lifetimes than PMOLED displays.

Figure 1.4(a) reflects the structure of a bottom-emission AMOLED display in which the light passes through the substrate [19]. Thus the substrate is limited to transparent materials, and the aperture ratio is diminished by the area lost to the pixel circuitry, resulting in a higher current density. Moreover, the aperture ratio becomes more critical when considering a more complex pixel

(a)

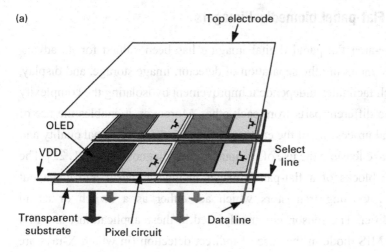

(b)

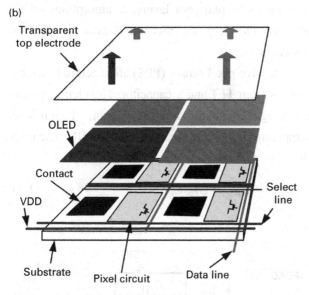

Figure 1.4 Bottom and top emission AMOLED pixel structure. Color versions of these figures are available online at www.cambridge.org/chajinathan.

circuit to compensate for both spatial and temporal non-uniformities. Hence, top-emission displays are preferred (see Figure 1.4(b)). It provides for more than an 80% aperture ratio, and the substrate is not required to be transparent [20].

1.2 Flat-panel biomedical imagers

Large-area flat-panel digital imaging has been valued for its advantages, including the separation of detector, image storage, and display, which facilitates independent improvement by isolating the complexity of the different parts from each other. Moreover, it enables the use of digital processing of the captured images to improve visual quality and to make feasible the use of computer-aided diagnostics [2, 21, 22]. The basic blocks of a flat-panel imager include a sensor and a readout circuitry using transistors which act either as a switch or as an amplifier. The sensor, commonly used in these applications, is a PIN or a MIS diode in the case of indirect detection (in which X-rays are converted to optical signals by phosphor layers) or amorphous selenium for direct detection (whereby the incident X-rays are directly converted to electrical charge).

Figure 1.5 shows a passive pixel sensor (PPS) architecture in which the pixel consists of a switch TFT and a capacitor. The charge generated by the sensor is integrated into the storage capacitor, which is read out by a charge-pump amplifier, while the switch TFT is ON. The gain of the PPS pixel is given as

$$V_{out} = Qt_{int}C_g. \tag{1.1}$$

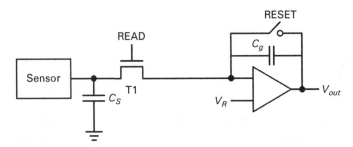

Figure 1.5 PPS imager pixel circuit adapted from [1, 2].

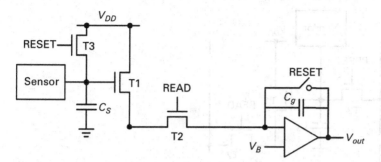

Figure 1.6 3-TFT APS imager pixel circuit adapted after [3].

Here, Q is the charge generated by the sensor and t_{int} the integration time. However, due to non-uniformities such as noise and leakage currents, the minimum level, detectable by PPS pixel, is limited.

To improve the sensitivity to small intensity signals, an active pixel sensor (APS) was introduced by Matsuura [3] (see Figure 1.6). Here, the storage capacitor is charged to a reset voltage. Then, the collected sensor charge into the storage capacitor modulates the current of the amplifier TFT (T1) as

$$I_{px} = g_m Q t_{int} \quad \text{and} \quad V_{out} = C_g I_{px} t_{read}, \tag{1.2}$$

in which g_m is the trans-conductance of T1, and t_{read} the time associated with the readout cycle. However, for high-intensity input signals, the on-pixel gain saturates the readout circuitry. In particular, for biomedical X-ray imaging applications, the significant contrast in the signal intensity of the different imaging modalities mandates unique pixel design.

Recently, hybrid mode pixel circuits have been reported that can operate between the passive and active readout for modalities employing high and low X-ray intensities, respectively. Figure 1.7 shows the 3-TFT hybrid pixel circuit presented in [23]. The issue with this type of a circuit is that it can be optimized for only active or passive operation. For example, in the active readout mode, a small storage pixel is required to improve the SNR [23], whereas, in the passive readout, a

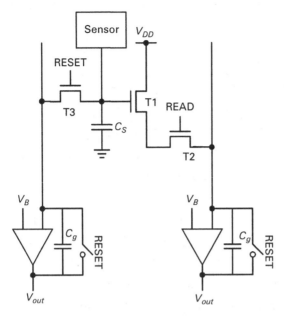

Figure 1.7 Hybrid active-passive imager pixel circuit adapted from [23].

large storage capacitor is needed to avoid saturation, improving the maximum detectable signal intensity. Also, a high-resolution (2-TFT) version of this pixel has been reported [24]. In addition to a limited dynamic range, this circuit suffers from cross talk and accelerated aging of TFTs. In particular, during the readout cycle of a row, the amp TFTs in pixels of inactivated rows are in the linear regime resulting in relatively high cross talk and accentuated aging [25]. Moreover, to achieve linear sensitivity, the amplifier TFT is biased in the linear regime, which makes it very susceptible to IR drop and ground bouncing.

1.3 Backplane technologies

The pixel circuits discussed above can be fabricated using different technologies, notably, poly silicon (poly-Si) [27, 28, 29] and hydrogenated amorphous silicon (a-Si:H) [3, 6, 30]. Poly-Si technology

Table 1.1 Comparison of TFT backplane technologies for large-area electronics [26].

Attribute	a-Si:H	Oxide	Poly-Si	mc/nc-Si:H	Organic
Circuit type	n-type	n-type	n-type/p-type	n-type/p-type	p-type
Mobility (cm^2/Vs)	< 1	~10	10~100	~1 to 10	<< 1
Temporal stability (ΔV_T)	issue	more stable than a-Si:H	more stable than a-Si:H	more stable than a-Si:H	improving
Initial uniformity	high	higher than poly-Si	low	potentially high	low
Manufacturability	mature	developing	developing	research	reseach
Cost	low	low	high	low	potentially low

offers high-mobility and complementary (n-type and p-type) TFTs [28, 29], but has an undesirable large range of mismatched parameters over an array [31, 32]. This is due to the random distribution of the grain boundary in the material [31].

In contrast, a-Si:H provides low mobility TFTs but does not provide p-type devices [33]. Also, the threshold voltage of TFTs increases (V_T-shift) under prolonged bias stress due to the inherent instability of a-Si:H material [34, 35]. Despite this, the technology provides good uniformity over a large area. More importantly, a-Si:H technology's industrial accessibility, by virtue of its usage in the AMLCD [7], provides for low-cost large-area electronics. In particular, an a-Si:H TFT backplane has the benefit of all the desirable attributes of the well-established a-Si:H technology, including low-temperature fabrication on plastic for eventual flexible electronics. Table 1.1 lists the attributes of different possible fabrication technologies.

In addition, promising research is being carried out on new materials such as hydrogenated nano/micro crystalline (nc/mc) silicon [36, 37, 38], organic semiconductors [39, 40], and more recently, the highly promising amorphous oxide semiconductors [41–43]. The nc/mc-Si:H and oxide semiconductor (e.g. indium gallium zinc oxide) technologies provide higher temporal stability (see [37, 38, 44–47] and references therein) and mobility (see [36, 43, 44, 47, 48] and references therein) compared to the ubiquitous a-Si:H technology. However, light-induced instability can be an issue [45, 46] requiring special driving techniques for threshold-voltage recovery [45, 47]. Despite this a variety of analog and digital circuits have been demonstrated [49, 50], including active matrix organic displays [51, 57] and imaging arrays [47, 52, 53]. On the other hand, an organic semiconductor has the potential for extremely low cost fabrication, including inkjet printing. However, this technology suffers from bias-induced [41, 42, 54–56] and environment-induced instabilities [43] and poor uniformity [44].

Regardless of the issues, the circuit solutions provided here apply to the broad range of materials families shown in Table 1.1 since they share the same problems of stability and (spatial and/or temporal) non-uniformity.

1.4 Organization

The challenges and design considerations for imaging and display applications are presented in Chapter 2, in which we discuss the principle of different driving schemes, including voltage and current programming. The major drawback of these driving schemes is settling time, which is investigated for both voltage and current programming; the proposed solution for improving voltage programming is presented in Appendix A.

A simple acceleration technique for current programming is detailed in Chapter 3. An imager/sensor pixel circuit is discussed, followed by short-term stressing to improve the temporal stability; a more detailed description of this technique is presented in Appendix B. Also, presented is a pixel circuit for implementation of a 16×12 biomedical sensor. To improve the pixel dynamic range, a variable capacitor is used to mitigate the integration of the charge generated by the sensor, and also to improve the gain for very low intensity input signals. The acceleration technique is also demonstrated for stabilizing AMOLED displays. Here, measurement and simulation results are presented for extreme spatial and temporal instability.

The use of a positive feedback to improve the settling time for very small programming currents (~100 nA) is discussed in Chapter 4. Since the system is non-linear, the stability of the system is investigated using the Lyapunov approach. The operation and stability of a circuit implementation of the driving scheme is also detailed. Measurement results for the circuit and the offset-leakage cancellation technique, developed for small current levels, are discussed.

For applications where cost is critical including small-area displays, a new charge-based compensation technique is presented in Chapter 5. Here, the cost of implementation is reduced by simplifying the external driver requirements. Moreover, a novel relaxation technique is introduced to improve the stability of the AMOLED pixel circuit. The implementation of 9-inch and 2-inch displays is discussed, followed by stability measurement results for the 9-inch displays. The charge-leakage technique is also adopted in the development of an extremely high dynamic range biomedical imager pixel circuit. The gain-adjustability measurement results of the circuit are presented and the noise performance of the pixel circuit is analyzed.

To meet the specification for high-resolution applications such as large-area AMOLED displays, a highly accurate compensation technique is required to reduce the effect of temporal/spatial mismatches to less than 0.5%. A successive calibration technique is proposed in Chapter 6. Its implementation using a single-bit current comparator is reviewed. Measurement results for the current comparator and successive calibration are discussed. This method can control the non-uniformities in other components such as sensor and OLED. Calibration of the OLED with this technique is described in Appendix B.

Finally Chapter 7 provides a summary along with key conclusions and recommendations.

2 Design considerations

We saw in Chapter 1 that technologies such as poly-Si, a-Si:H, and organic semiconductors are available for the fabrication of pixel circuits. Figure 2.1 demonstrates the three most used TFT structures. Since the bi-layer staggered bottom-gate structure requires fewer mask sets and processing steps, it is highly adopted in industrial scaled a-Si:H fabrication. However, this structure is prone to a higher leakage current, since the back-side of the a-Si:H layer is damaged during the process. An alternative solution to this structure is tri-layer structure in which an etch stopper layer is used to preserve the a-Si:H layer. However, tri-layer structure has more mask layers and process steps compared to bi-layer structure which makes the industry reluctant to adopt it. For poly-Si TFTs, the coplanar top-gate structure is the most common structure. This structure enables self-alignment, resulting in smaller design rules and TFT sizes.

2.1 Temporal and spatial non-uniformity

Each of these fabrication technologies is associated with drawbacks for circuit design. However, the key challenge in using the available technologies is the temporal or spatial non-uniformity. In a-Si:H and oxide technologies, the threshold voltage of the TFTs tends to shift (V_T-shift) under prolonged bias stress condition (denoted in Figure 2.2). Considering that each pixel in most applications experiences different biasing conditions, the V_T-shift will increase the non-uniformity across the panel over time. This phenomenon occurs due to charge trapping and/or defect

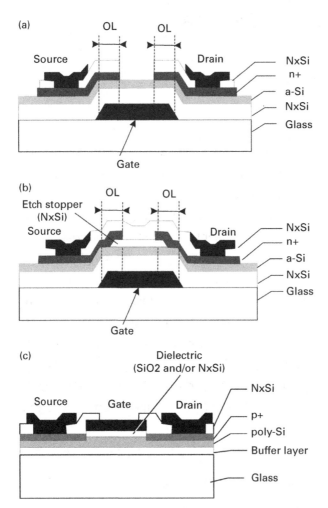

Figure 2.1 Different TFT structures: (a) bi-layer inverted staggered, (b) tri-layer inverted staggered, and (c) coplanar.

state creation [58, 59]. The V_T-shift has been modeled under different conditions including constant voltage [58, 59], constant current [60], and pulsed stress conditions [61, 62]. Depending on different applications, one of these models can be applied to extract the aging of the pixel. However, in the applications that TFT is under a constant current

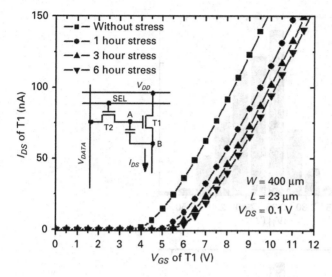

Figure 2.2 Biased induced V_T-shift (stress condition: $V_{GS} = 10$ V, $V_{DS} = 0.1$ V).

stress, the V_T-shift is severe [60] and unlike the TFT under constant voltage stress, the V_T-shift tends to increase forever.

Also, poly-Si TFTs are more stable but suffer from initial non-uniformities caused by recrystallization methods [31, 32]. Since the channel of a TFT consists of several randomly oriented crystalline grains, the interface of these grains (grain boundaries) can manipulate the mobility and V_T. Figure 2.3 shows the stochastic results gathered from a set of 600 poly-Si TFTs. It is evident that both V_T and mobility are prone to mismatch. Similarly oxide TFTs suffer from non-uniformities especially from long-range non-uniformity [41–53].

2.2 Compensation schemes

Although the initial spatial mismatches result in an after-fabrication non-uniformity reducing the yield significantly, the temporal instability increases the non-uniformity over time, and

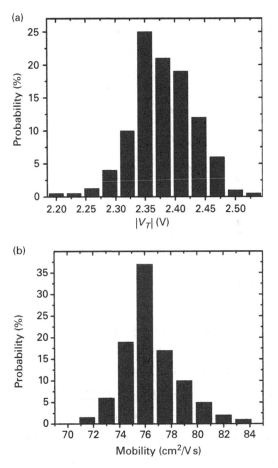

Figure 2.3 (a) V_T and (b) mobility variation in poly-Si TFTs (adapted from [32]).

shortens the lifetime of the device. However, the initial spatial mismatches and temporal instability tend to cause the same effect in the pixel circuit. As a result, most of the compensation techniques control both spatial and temporal non-uniformity. The two primary compensation techniques, introduced to improve the yield/lifetime and panel quality, are current and voltage driving schemes [6].

2.2.1 Current driving scheme

Current-programmed active matrix (AM) architectures are attractive for displays and sensors independent of the fabrication technology because of their ability to tolerate mismatches and non-uniformity caused by aging. Figure 2.4 illustrates two different pixel circuits, based on the current cell and current mirror. Two current-programmed pixel circuits (CPPCs) for AMOLED displays and flat-panel imagers are depicted in Figure 2.5. Here, a shared data line is connected to the I_{data} port of the pixels in one column and a current source is used as part of peripheral circuitry to program the pixels row by row. Figure 2.6(a) shows a detailed model of the data line, in which R_i ($i = 1$ to n) stems from the sheet resistance of the metal, C_i ($i = 1$ to n) the parasitic capacitances stemming from the line and pixels, and I_{Li} ($i = 1$ to n) the leakage current contribution of the ith pixel. The sources of parasitic capacitances are fringing capacitance between long trances, overlap between vertical horizontal signals, switches' overlap and gate-source capacitances, and the signal line with the common blanket electrodes. While one can reduce the parasitic capacitance by methods

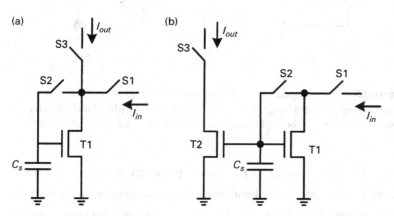

Figure 2.4 Current-programmed pixel circuits; (a) current cell and (b) current mirror.

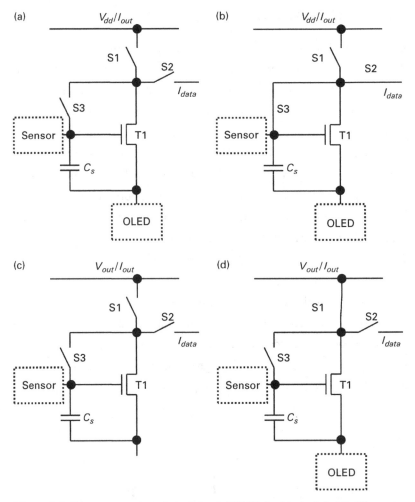

Figure 2.5 Current-programmed pixel circuit (CPPC) for (a) pixel circuit, (b) reset/
programming cycle, (c) integration cycle (only for sensor applications), (d) driving/
readout cycle (driving for displays and readout for sensors; adapted from [63, 64]).

such as self-aligned TFTs, extra passivation layers between
common electrodes and signal lines, step-coverage between vertical
and horizontal crossing signals, the value of the parasitic capacitance will stay substantial due to the long signal paths. In addition,

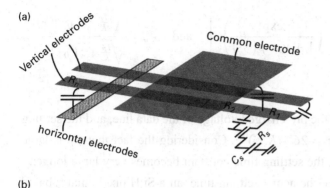

(a)

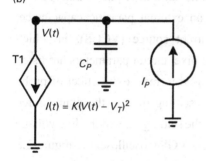

(b)

Figure 2.6 Line model for a column during the programming cycle (a) and its simplified equivalent (b) adapted from [99].

the switches, used to form the pixel circuits, are not ideal, adding resistance (R_s) to the path of the programming current. It is noteworthy that the value of R_i ($i = 1$ to n) is a few ohms whereas the R_s can be as high as a few 100 kΩ. To identify the major source involved in the settling behavior of a CPPC, the first-order model (Figure 2.6(b)) is used for the analysis. Here, C_P, $K(V(t) - V_T)^2$ represent the parasitic capacitance of the line and first-order model of the drive TFT (no channel length modulation and contact resistance), respectively. Here, $V(t)$ is the time-dependent voltage of the line as a result of applied programming current (I_P) and K the I–V gain. As a result, the pixel dynamic during the programming can be written as

$$V(t) = \left[\sqrt{\frac{I_P}{K}} \cdot \left(\frac{1 - v_a \exp\left(-\frac{t}{\tau}\right)}{1 + v_a \exp\left(-\frac{t}{\tau}\right)} \right)^2 \right] \quad \text{and} \quad V_a = \frac{\sqrt{\frac{I_P}{K}} - (V_0 - V_T)}{\sqrt{\frac{I_P}{K}} + (V_0 - V_T)}.$$

$$(2.1)$$

In which, V_0 is the pre-charged voltage of the data line, and the settling time constant $\tau = 2C_P/(K.I_P)^{0.5}$. Considering the fact that C_P is large and K is small, the settling time constant becomes very large longer.

To investigate the actual settling time, an a-Si:H pixel circuit, based on Figure 2.4(b), is used but with an external parasitic capacitance, along with a voltage-controlled current source (VCCS), in which $I_{out} = V_{in}/R1$ (see Figure 2.7(a)). The pixel circuit parameters are listed in Table 2.1. A unity gain buffer is used to reduce the effect of probe loading on the critical node. The 5% settling time of the drain-source current (I_{DS}) of T1 which is similar to the settling time of the line voltage is measured by a Tektronix TDS 5054 5 GS/s oscilloscope. Figure 2.8 indicates that the settling time increases linearly with parasitic capacitance (C_P) as predicted in (2.1). As shown in Figure 2.8, the settling time is higher than 70 μs and becomes even more severe for larger parasitic capacitances. Generally, the row time is reduced for larger panels at the same time as the parasitic capacitance increases due to longer data lines and more rows. For example, for a typical small display with 240-row with a 60-Hz frame rate, the programming time for each row is less than 70 μs and its corresponding parasitic capacitance is in the range of 20 pF, whereas for a large TV full HD with 1080-row at 120-Hz frame rate, the programming time is around 7 μs and parasitic capacitance is of the range of nF. Consequently, using pure current programming does not work in any of the possible applications.

The effect of the pre-charging voltage (V_0) on the settling time is demonstrated in Figure 2.9. As the pre-charging voltage gets closer to the final voltage (V_f) characteristic for a specific programming current,

Table 2.1 *Parameters of the amorphous silicon current cell.*

Name	Description	Value
W/L(T1)	Aspect ratio of T1 (μm)	400/23
W/L(T2)	Aspect ratio of T2 (μm)	100/23
W/L(T3)	Aspect ratio of TFT used for OLED (μm)	100/23
C_S	Storage capacitance (pF)	1

(a)

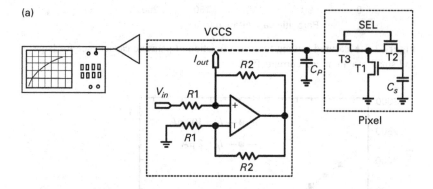

(b)

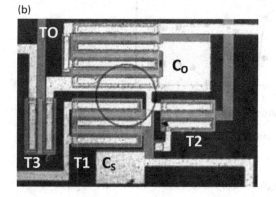

Figure 2.7 (a) Discrete VCCS and (b) photomicrograph of fabricated current cell in amorphous silicon technology [99].

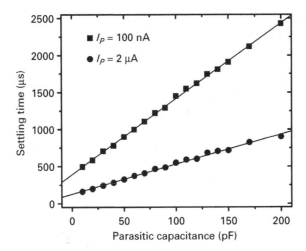

Figure 2.8 Measured settling time as a function of parasitic capacitance for large and small programming currents [99].

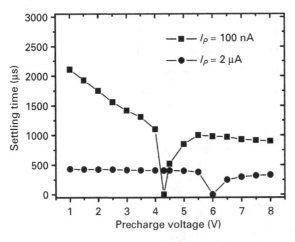

Figure 2.9 As in Figure 2.8, but settling time is as a function of the pre-charge voltage [99].

the settling time drops significantly. However, due to the high level of mismatch, a priori selection of a suitable pre-charging voltage is not practical. Figure 2.10 shows the effect of programming current (I_P) on the settling time, where the $V_0 - V_f$ is 1 V. As (2.1) predicts, the settling

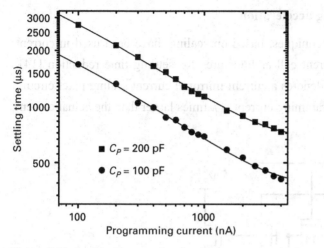

Figure 2.10 Settling time as a function of the programming current for different parasitic capacitances [99].

time should drop linearly on a logarithmic scale as the programming current increases. However, the measurement results do not show the presence of perfect correlation at large current levels. This could be due to the resistance of the switch transistors. Assuming that the V_{DS} of the switches is small, the resistance of the switches is given by

$$R_S \approx \frac{1}{K_S(V_H - V_L - V_{TS})}. \tag{2.2}$$

Here, K_S is the gain in the I–V characteristics of switches, V_H the select voltage, V_L the voltage on the data line, and V_{TS} the threshold voltage of switches. Thus, for a given select voltage, the resistance of the switches increases as the voltage of the data line increases due to larger currents. As a result, the actual settling time deviates more than that predicted by the first-order model shown in Figure 2.6(b).

Although, several solutions have been proposed to accelerate the settling time [14, 66–68], the effectiveness of those solutions, coupled with power consumption and/or circuit overhead, is a lingering issue.

2.2.1.1 Scaling acceleration

Acceleration techniques, based on scaling, have been used in current mirror and current cell architectures for settling time reduction [14]. Figure 2.11(a) denotes a current-mirrored current-scaling pixel circuit. Here, the programming current is k times larger than the actual current

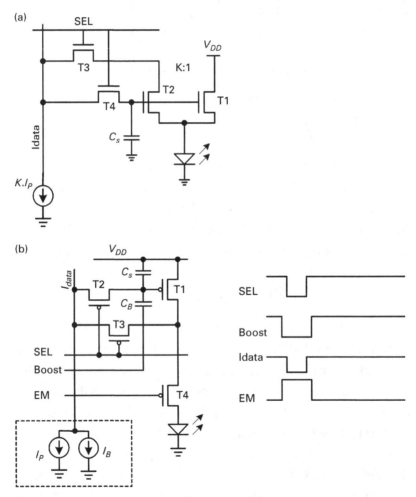

Figure 2.11 Acceleration driving scheme: (a) scaling (adapted from [14]) and (b) additive (adapted from [66]).

required for the pixel circuit and the current is scaled down at the pixels. In particular, for very low currents, of the order of 100 nA, k should be several hundred.

2.2.1.2 Additive acceleration

The other solution is additive acceleration in which bias current (I_B) is added to the actual current and is subtracted locally at the pixel [66–68] (illustrated in Figure 2.11(b)). This method is more effective for low current levels compared with the scaling method, and is applicable to both current mirror and current cell pixel architectures [17]. However, the bias current can be large, leading to high power consumption. This can be a major concern in portable devices such as small displays, and more importantly, the subtraction of this bias current potentially reduces the immunity to mismatches and temperature fluctuations.

2.2.2 Voltage driving scheme

The voltage driving schemes have been mostly adapted in the AMOLED pixel circuits [6, 69]. In voltage driving schemes in order to compensate for the V_T-shift in a-Si:H or V_T-mismatch in poly-Si TFTs, the gate-source voltage (V_{GS}) of drive TFT must include the programming voltage and the V_T of the drive TFT. Here, the major operating cycles are V_{comp}-generation, V_T-generation, programming, and driving [70, 71]. In the pre-charging cycle, a compensating voltage is stored in the storage capacitor. During the V_T-generation cycle, the voltage stored in the storage capacitor discharges through the diode-connected drive TFT until it turns off, so the gate-source voltage is equal to the V_T of the drive TFT. In the current-regulation cycle, a programming voltage (V_P) is added to the generated V_T, resulting in a gate-source voltage as $V_P + V_T$. Therefore, during the driving cycle, the pixel current is given by

$$I_{pixel} = K(V_P)^{\alpha}. \qquad (2.3)$$

Here, K is the gain parameter in the TFT I–V characteristics [65].

Based on the method used to add a programming voltage (V_P) to the generated V_T, the voltage-programmed pixel circuits (VPPCs) can be divided into four different categories: stacked, parallel-compensation, bootstrapping, and mirror VPPCs.

2.2.2.1 Stacked voltage programming

Figure 2.12 shows the simplified circuits for a typical stacked VPPC during different operating cycles [72, 73]. Here, C_S is the storage capacitor, and C_{OLED} the OLED capacitor. During the pre-charge cycle, node B is charged to $-V_{comp}$. During the V_T-generation cycle, node B is discharged until T1 turns OFF so that the voltage at node B becomes $-V_T$ of T1. In the current-regulation cycle, node A is charged to V_P. Considering that C_{OLED} is large, the voltage at node B stays at $-V_T$, resulting in the V_{GS} of T1 as $V_P + V_T$.

The pixel circuit, based on this driving scheme, is depicted in Figure 2.13. In the 2-TFT pixel circuit [73], V_{dd} line is used to

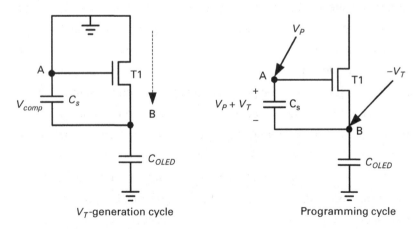

Figure 2.12 Stacked VPPC configurations during different operating cycles.

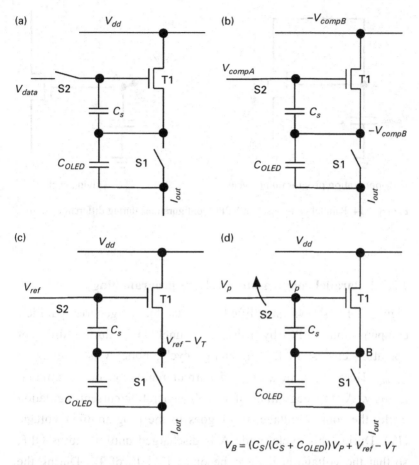

Figure 2.13 Stacked VPPCs; (a) pixel circuit, (b) initialization cycle, (c) V_T-generation cycle, and (d) programming cycle (S2 turns for short time to avoid losing charge at node B, adapted from [73]).

discharge the voltage at node B during initialization cycle. During V_T-generation cycle, node B is charged back to $V_{ref} - V_T$. Also, the programming voltage is written into storage capacitor (C_S) through S2. Here, the OLED can be used as S1. Since T1 will turn on as the pixel is being programmed, it can charge node B to the point of losing the compensation value.

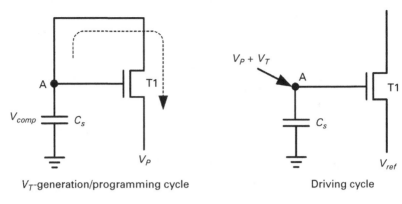

V_T-generation/programming cycle   Driving cycle

Figure 2.14 Parallel-compensation VPPC configurations during different operating cycles.

2.2.2.2 Parallel-compensation voltage programming

Figure 2.14 shows simplified circuits for a generic parallel-compensation VPPC by using n-channel TFTs during different operating cycles. In the pre-charge cycle, node A is charged to V_{comp}. Here, V_T generation and current regulation occur simultaneously. At the beginning of the V_T-generation/current-regulation cycle, the source voltage of T1 goes to the programming voltage (V_P). During this cycle, node A is discharged until T1 turns OFF, so that the voltage at node A becomes $V_P + V_T$ of T1. During the driving cycle, source voltage of T1 goes to V_{ref}, such that the V_{GS} becomes $V_P + V_T - V_{ref}$.

Figure 2.15 demonstrates a parallel-compensation VPPC during different operating cycles [74, 75]. Here, the compensation and programming occur in the second operating cycle. The main challenge is the voltage at source of T1 is different during programming and driving cycles. As a result, the pixel is sensitive to the voltage variation at the source of T1 due to aging, ground bouncing, or other phenomena.

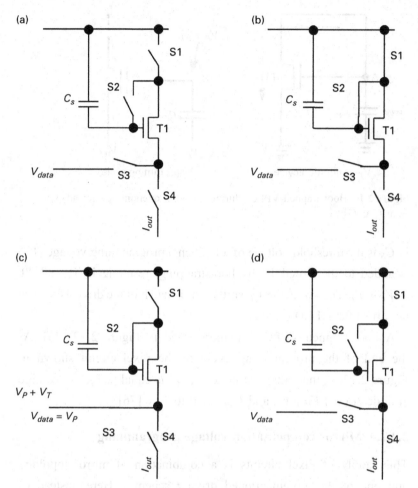

Figure 2.15 Parallel-compensation VPPCs: (a) pixel circuit, (b) initialization cycle, (c) V_T-generation/programming cycle, (d) driving cycle; circuit (adapted from [74, 75]).

2.2.2.3 Bootstrapping voltage programming

Figure 2.16 shows a typical bootstrapped VPPC by using n-channel TFTs during different operating cycles [6]. During the V_T-generation cycle, the voltage at node A (V_{comp}) is discharged through the diode connected drive TFT (T1) until T1 turns off. Thus, the voltage stored

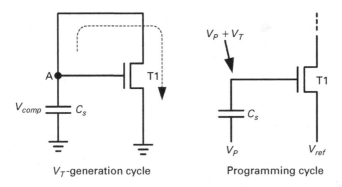

V_T-generation cycle   Programming cycle

Figure 2.16 Bootstrapped VPPCs during different operating cycles adapted from [76, 77].

in C_S is the threshold voltage of T1. Then a programming voltage (V_P) is added to the stored V_T by bootstrapping, resulting in V_{GS} of T1 as $V_P + V_T$. Therefore, the V_T-shift/V_T-mismatch of the drive TFT does not affect the OLED current.

A bootstrapped VPPC is demonstrated in Figure 2.17 [76]. At the end of the programming cycle of the pixel circuit shown in Figure 2.17(a), the voltage at node C (V_C) is equal to V_P, the voltage at node A (V_A) $V_P + V_T$, and V_{ref} equal to V_{DD} [76].

2.2.2.4 Mirror compensation voltage programming

This family of pixel circuits is a combination of mirror topology and one of the aforementioned driving schemes. Here, instead of compensating the V_T-shift/V_T-mismatch in the drive TFT, the V_T-shift/V_T-mismatch in the mirror TFT is compensated. In poly-Si technology, the principal assumption for these circuits is that the short-range mismatch is negligible. However, in a-Si:H technology, the drive and mirror TFTs must have the same biasing conditions in order to have the same V_T-shift.

Two mirror VPPCs, based on parallel-compensation, are shown in Figure 2.18. In the pixel circuit of Figure 2.18(a), the programming

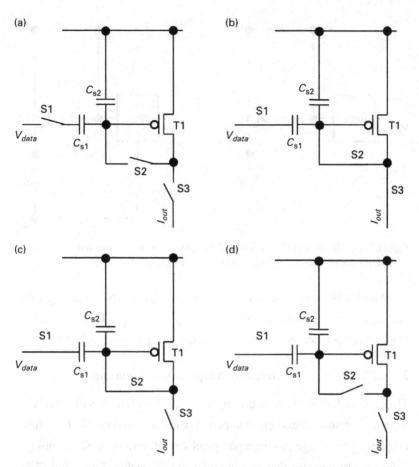

Figure 2.17 Bootstrapped VPPCs; (a) pixel circuit, (b) initialization cycle, V_T-generation cycle, and (c) programming cycle (during driving all the switches are OFF and only S3 is ON; adapted from [75]).

block is isolated from the main current path. Therefore, the circuit can be simpler. However, this does not consider the effect of voltage at I_{out} terminal during programming. Since this voltage may change due to different effects (e.g. ground bouncing, aging, temperature, …) the pixel current may suffer from it during the driving cycle. In the next circuit, however, the programming and driving current paths are the

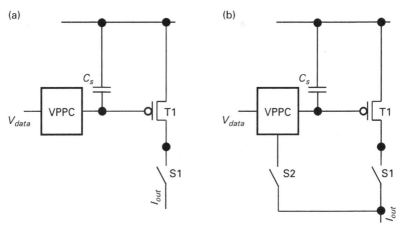

Figure 2.18 Mirror VPPCs; (a) isolated programming block (adapted from [78]), (b) alternating programming block (adapted from [79]).

same and it alternates between them through S1 and S2. Although, the circuit can become more complex, the pixel current does not suffer from the difference between programming and driving voltages.

2.2.2.5 Spatial mismatch and temperature variation

The drawback of the voltage driving scheme is the high level of sensitivity to spatial mismatch and environmental parameter variations. Therefore, employing a voltage-programmed pixel circuit with poly-Si technology is an area of concern due to a large spatial mismatch. Following (2.3), K is a function of device geometry and mobility. Therefore, any change in the geometry due to spatial mismatch directly affects the pixel current. Also, since the TFT mobility is a strong function of temperature, any temperature change results in pixel current variation.

However, the stacked voltage-programmed pixel circuits are less sensitive to mismatch and temperature variation. In the pixel circuit shown in Figure 2.13(b), since T1 is ON during the third operating cycle, the stored gate-source voltage of T1 reduces. The V_{GS} of T1 can be written as [80]

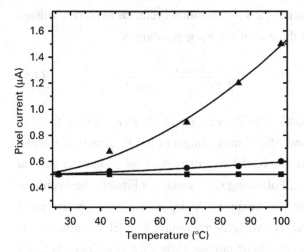

Figure 2.19 Temperature stability measurement results for stacked VPPC (circles for $C_{OLED} = 6$ pF and squares for $C_{OLED} = 8$ pF) and conventional (triangles) driving schemes [80].

$$V_{GS} \approx V_P \exp\left(-\frac{K\tau_{CR}}{C_S + C_{OLED}}\right) + V_T, \qquad (2.4)$$

where τ_{CR} is the timing budget of the current-regulation cycle. This indicates that the stored V_{GS} of T1 depends on K. Furthermore, a change in K due to spatial mismatch, temperature variation, and mechanical stress changes the stored V_{GS} of T1 in the reverse direction. Although, the current of T1 depends on both V_{GS} and K, this reverse variation makes the pixel less sensitive to mismatch. Figure 2.19 demonstrates the pixel current for the stacked VPPC and conventional 2-TFT driving schemes. The pixel current surpasses close to 300% for the latter after 70 °C, whereas the current changes less than 40% for the compensation driving scheme [80].

2.2.2.6 Imperfect compensation

The principal obstacle in using voltage-programmed pixel circuits in large-area devices is the imperfect compensation during the V_T-generation cycle [70, 71]. Considering that the drive TFT is in the saturation region during the V_T-generation cycle, the overdrive voltage of the drive TFT at the end of the V_T-generation is

$$V_{OV}(\tau_{GC}) = \frac{V_{comp} - V_T}{\frac{K}{C_T}(V_{comp} - V_T)\tau + 1}, \tag{2.5}$$

where C_T is the total capacitance that is effective during the V_T-generation cycle, and τ the timing budget of the V_T-generation cycle. For perfect compensation, the overdrive voltage should be zero at the end of this cycle. Following (2.5) since τ is limited, the overdrive voltage is not zero, inducing a V_T-dependent error in the pixel current. In the stacked voltage-programmed pixel circuits, C_T is $C_S + C_{OLED}$, and in other pixel circuits C_T is C_S. Since C_{OLED} is larger than the storage capacitor (C_S), the imperfect compensation can be more severe in the stacked voltage-programmed pixel circuit. Measurement results for the difference between threshold voltage (V_T) and generated threshold voltage (V_{TG}) with different timing budgets assigned to the V_T-generation cycle are shown in Figure 2.20. While

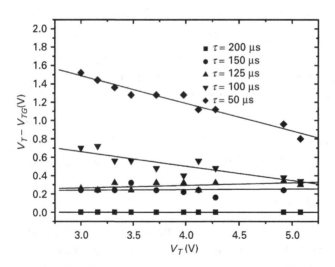

Figure 2.20 Effect of limited timing budget on V_T generation [71].

V_{TG} and V_T are exactly the same for a 200-μs V_T-generation, V_{TG} has a fixed error for medium V_T-generation cycles ($\tau = 150$ μs and 125 μs), becoming more severe for short V_T-generation cycles ($\tau = < 125$ μs). Consequently, the errors in the generated V_T (V_{TG}) for small timing budgets preclude the use of voltage compensating techniques in AMOLED displays. The experiments are carried out by using a TFT with an aspect ratio 400 μm/23 μm.

2.3 Design considerations for AMOLED displays

In addition to current and voltage driving schemes, several other driving schemes have been proposed for AMOLED displays including optical feedback [81], electrical (current or voltage) feedback [82, 83], and digital [84] (time-based) driving schemes. In optical feedback, a photo diode/TFT is used to monitor the OLED luminance, and adjust the gate voltage of the drive TFT accordingly [81]. Thus, the optical feedback can theoretically compensate for all the undesirable effects such as V_T shift/mismatch, temperature variation, and OLED aging. However, the problem with this driving scheme is the instability of the sensor, high susceptibility to cross talk, and complex pixel circuits. On the other hand, electrical feedback has more stable operation, but comes with the cost of higher cost driver and lower resolution pixel circuit. Also, despite the simplicity of the digital driving schemes [84], they suffer from the contrast ratio due to missing low gray scales. Also, the number of gray scales is limited in this type of driving scheme.

In order to design a suitable driving scheme for different AMOLED displays, one needs to know the major design considerations which can be listed as lifetime, differential aging and mura, power consumption, aperture ratio, IR drop, and implementation cost.

2.3.1 Lifetime and yield

Display lifetime is when the display luminance has dropped to half of its initial value. This occurs due to OLED luminance degradation and TFT degradation. For simplicity, it is assumed that in a-Si:H AMOLED display, TFT is the only source of aging and the compensation scheme can perfectly manage the effect of aging. However, the compensation is limited to the given headroom between maximum overdrive voltage and the operating voltage of the driver. To find out the limitation due to the operating voltage, the V_T-shift model for constant current is used. The shift in threshold voltage under constant current is given by [60]

$$\Delta V_T = \frac{(I_{DS}/K)^{\frac{\gamma}{\alpha}}}{\left(1 + \frac{1}{\alpha}\right)^\gamma} \left(\frac{t}{\tau}\right)^\beta, \qquad (2.6)$$

where τ, β, γ are process/device dependent parameters [60]. The time required to reach some maximum allowable level can be expressed as

$$t = \tau \left(\frac{\Delta V_{Tmax}\left(1 + \frac{1}{\alpha}\right)^\gamma}{(I_{DS}/K)^{\frac{\gamma}{\alpha}}}\right)^{\frac{1}{\beta}}. \qquad (2.7)$$

For convenience, the current–voltage characteristic of the drive TFT is written, in the following form, by assuming operation is in the saturation regime

$$I_{DS} = K(V_{GS} - V_T)^\alpha, \qquad (2.8)$$

where $K \propto (W/L)\mu$ where μ is mobility.

$$V_{GSmax} - V_{Tmax} = (I_{DS}/K)^{\frac{1}{\alpha}}, \qquad (2.9)$$

where

$$V_{Tmax} = V_{T0} + \Delta V_{Tmax}. \qquad (2.10)$$

Here, V_{T0} is the initial threshold voltage. Failure occurs when V_{GSmax} reaches $V_{DD} - V_{OLED}$ ($\equiv V_{dd}$) then the maximum allowable V_T-shift is given as

$$\Delta V_{Tmax} = V_{dd} - V_{T0} - (I_{DS}/K)^{\frac{1}{\alpha}}. \qquad (2.11)$$

Therefore the lifetime following (2.7) can be estimated as

$$t_{lifetime} = \tau \left(1 + \frac{1}{\alpha}\right)^{\frac{\gamma}{\beta}} \left((V_{dd} - V_{T0})\left(\frac{\mu_{FE}C_iW}{2LI_{DS}}\right)^{\frac{\gamma}{\alpha}} - \left(\frac{\mu_{FE}C_iW}{2LI_{DS}}\right)^{\frac{\gamma-1}{\alpha}}\right)^{\frac{1}{\beta}}.$$

$$(2.12)$$

Following (2.12), the maximum allowable V_T-shift for a given V_{dd} increases with the larger size of drive TFT. However, the size of drive TFT is limited by aperture ratio and pixel size. The required lifetime varies in different applications. For a smaller display with less area for large drive TFT, the lifetime is also smaller (~3000 hours). On the other hand, for larger displays, the required lifetime is around 50 000 hours. As is explained in Chapter 5, the size of drive TFT is limited by other factors such as the OLED current density. Thus, achieving such a lifetime without suppressing the aging can be challenging.

Although poly-Si backplane is more stable, the level of mismatch, which can be compensated, is limited to the V_{dd} and maximum required current, resulting in limited yield.

2.3.2 Differential aging and mura

Due to different non-idealities such as charge injection (see Chapter 6), the compensation techniques are not perfect. As a result, after compensation, the luminance difference across the panel may increase which is called differential aging (for temporal non-uniformity) or mura (for spatial non-uniformity). The amount of acceptable differential aging (or mura) changes, based on applications. For example, the amount of differential aging for mobile application is approximately 2% after aging the display with a white-and-black checker board for 120-hour. On the other hand, it should be less than 0.5% for a display intended for TV application.

2.3.3 Power consumption

Power consumption in a display consists of two distinct parts: panel and the driver. Also, the power consumption of an AMOLED panel stems from the programming and the driving power consumption. The power consumption, during driving cycle, is mostly due to charging and discharging different parasitic capacitors, particularly in VPPCs. The power consumption, during driving cycle, stems from the current passing through the OLED and drive TFT. Thus for a given OLED, to reduce the power consumption of the panel, the voltage drop across the TFT needs to be reduced which is limited by the size of TFT, required brightness, and required lifetime. However, as a rule of thumb, the number of TFTs, within the active path during the driving cycle, should be as low as possible (e.g. one or two) to reduce the power dissipation.

2.3.4 Aperture ratio

The aperture ratio is the area of the OLED to the total area of the pixel. Since the OLED degradation is a function of the current density [10, 81, 85], for a given brightness (resulting in a required current), the OLED lifetime increases as the aperture ratio increases. To improve the aperture ratio, fewer TFTs in a pixel and improved layout are required.

2.3.5 IR drop and ground bouncing

Although the current level for each individual pixel is low (around few µA), the total current which passes through the common electrode can be significant due to the high number of pixels in a display. As a result, the effective ground (or V_{DD}) voltage can be different for each pixel resulting in luminance gradient across the panel. This effect is simulated using finite element method for different structures (denoted in Figure 2.21). It is clear that by connecting more sides of the common electrode to the voltage source, the voltage drop can be reduced and the voltage gradient becomes smoother. However, the pixel circuit and driving scheme should be able to tolerate the ground bouncing and IR drop, since they change as the current density varies with different pictures shown on the display.

2.3.6 Implementation cost

Another important aspect of the design is the cost, in particular for small-area displays for portable applications. In a display structure, the cost is enforced by yield and driver components. To improve the yield, a more stable pixel circuit with fewer TFTs is required. In addition, the driving scheme should not increase the complexity of the drivers. In particular, the number of controlling or data signals required for each column and row should be reduced. For example, if a driving scheme needs a data line and a monitor line for each column, the number of source driver pads increases dramatically. Considering that the source drive is mostly pad-limited, two lines per column doubles the size of the driver leading to a higher cost.

2.4 Design considerations for flat-panel imager

Most of the discussions in the previous section about lifetime, uniformity, and power consumption can be applied to the imaging application as well. However, some design considerations are more

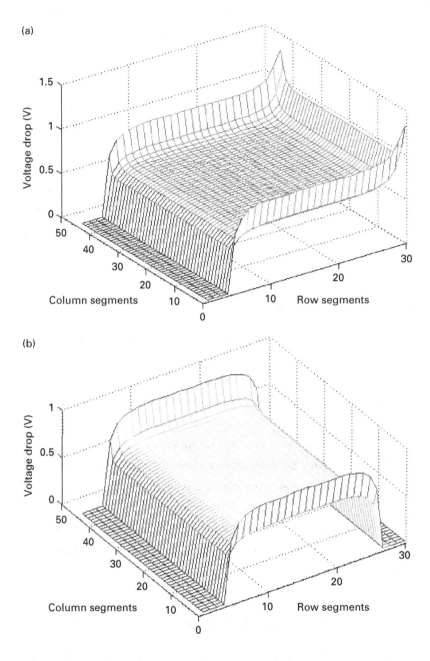

Figure 2.21 Voltage gradient across the common electrode based on different sides connected to the voltage source: (a) one side, (b) two sides, (c) three sides and (d) four sides. Color versions of these figures are available online at www.cambridge.org/chajinathan.

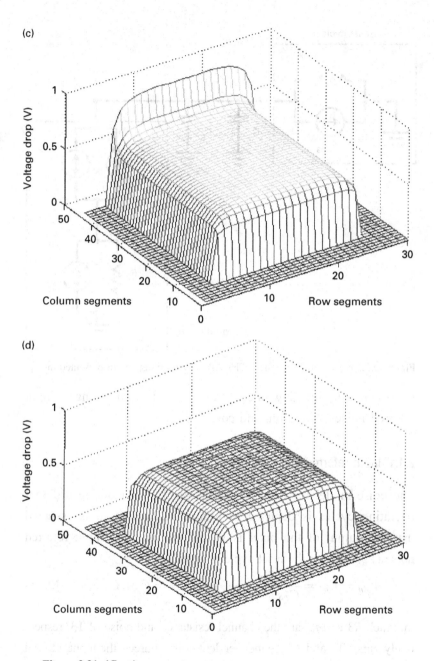

Figure 2.21 (*Cont.*)

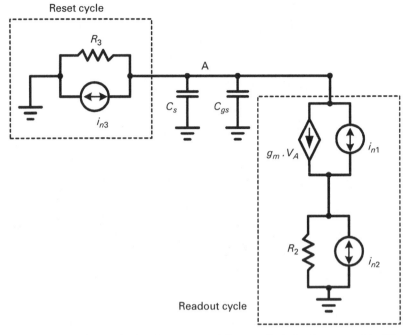

Figure 2.22 Noise model for the 3-TFT APS imager pixel circuit presented in [3].

important for imaging including input dynamic range, input referred noise, resolution, and cost.

2.4.1 Input referred noise and dynamic range

The noise model for the 3-TFT APS pixel circuit during different operating cycles is depicted in Figure 2.22. Here, it is assumed that the output of the pixel is virtual ground. The input referred noise (V_{Rn}) during the reset cycle is given by

$$V_{Rn} = V_{n3} = \frac{i_{n3}}{C_T s + 1/R3} \quad \text{and} \quad C_T = C_s + C_{gs}, \qquad (2.13)$$

in which $R3$ and i_{n3} are the channel resistance and noise of T3, respectively. Since T2 and T1 are independent noise sources, the input referred noise associated with the readout cycle is given by

$$V_{Rdn1} = \frac{i_{n1}}{C_T s + g_m} \quad \text{and} \quad V_{Rdn2} = \frac{i_{n2}}{C_T s + g_m}. \tag{2.14}$$

The total input referred noise can be given as:

$$\begin{aligned} V_{n-in} &= \left(|V_{Rdn1}|^2 + |V_{Rdn1}|^2 + |V_{Rn}|^2 \right)^{0.5} \\ &= \left(\frac{i_{n1}^2 + i_{n2}^2}{C_T^2 \omega^2 + g_m^2} + \frac{i_{n3}^2}{C_T^2 \omega^2 + 1/R_3^2} \right)^{0.5}. \end{aligned} \tag{2.15}$$

The low end of a dynamic range is defined by the maximum input referred noise of the system.

$$q_{min} > (q_n + q_s) \quad \text{and} \quad q_{min} \propto \frac{I_{min}}{\eta} \tag{2.16}$$

in which, I_{min} is the minimum input signal intensity, η the conversion efficiency of the sensor, q_s the noise associated with the sensor, q_n the total input referred noise in terms of number of electrons. Following (2.16) and (2.17), the sensitivity of the pixel sensor is limited by input referred noise and conversion efficiency of the sensor. On the other hand, the high end of input signal is determined by the dynamic range of readout circuitries.

$$q_{max} = \frac{V_{HS}}{A_T} \quad \text{and} \quad q_{max} \propto \frac{I_{max}}{\eta}. \tag{2.17}$$

Here, A_T is the total system gain and V_{HS} is the maximum output voltage of the readout circuit. Considering using a charge-pump amplifier as readout circuitry with a capacitance of C_g, A_T can be written as

$$A_T \propto \frac{C_S g_m t_{Rd}}{C_g} \tag{2.18}$$

where g_m is the trans-conductance of the amplifier TFT in case of active mode. Also, t_{Rd} is the readout time. Thus to increase the high end of the dynamic range one can use a larger C_g or smaller C_S.

2.4.2 Implementation cost

Due to the need for a low volume of biomedical imagers, their cost is very high. A multi-modal platform can improve the overall cost by sharing the device between several applications. However, due to the high contrast between the dynamic ranges of the different applications, the multi-modal platform must be able to handle a wide dynamic range.

2.5 Summary

Spatial and temporal uniformity is the major concern in development of AMOLED displays and flat-panel imagers. Although current programming can improve the uniformity and reduce the pixel sensitivity to device variation, it suffers from long settling time. Voltage programming, however, is faster but it is prone to imperfect compensation. To implement stable and uniform devices using the existing technologies, new compensation techniques, improving the uniformity, lifetime, and yield, are required.

Also, the new driving scheme should provide for low power consumption, low implementation cost, and high resolution for AMOLED display. In addition to low cost, low power consumption, the proposed driving schemes should result in improving the dynamic range and lowering the effect of input referred noise.

As a result, the design strategy has been based on the fact that stability and uniformity have the highest priority. Seeking an optimized solution for each application instead of a universal solution has been the other principle of the design strategy.

Copyright notices

Figure 2.7 adapted from [99].

Figures 2.8, 2.9, and 2.10 © 2009 IEEE. Reprinted, with permission, from [99].

Figure 2.20 © 2008 IEEE. Reprinted, with permission, from [70].

Portions of the text are © 2008, 2009 IEEE. Reprinted, with permission, from [99] and [70].

3 Hybrid voltage–current programming

This chapter presents accelerated programming/resetting of active pixel circuits based on bias currents presented in [87]. The settling time in current programming depends strongly on the V_T of the drive transistor and initial line voltage (V_0). Figure 3.1 shows the settling time of the pixel circuit depicted in Figure 2.5(a) as a function of the initial voltage. It is evident that the settling time changes as the initial voltage changes. As a result, since the voltage of the previous pixel remains on the data line, the programming of the new pixel is affected. Pre-charging the data lines to a specific voltage can control the effect of the initial voltage, but this can increase the power consumption considerably. Moreover, initial voltage is a function of the V_T. Thus, the settling time cannot be managed by a fixed pre-charging voltage, because the value of V_T is varying.

Thus, a dynamic line pre-charging scheme, coupled with a large current, is required to improve the settling time. To develop this driving scheme, a fix current is required to program the pixels so that the initial voltage becomes independent of the pixel content. Also, in-pixel current reduction/division is used to adjust the pixel current accordingly. This driving scheme is called the current-biased voltage-programmed (CBVP) scheme [86, 87].

3.1 Multi-modal biomedical imaging pixel circuit

Figure 3.2(a) demonstrates multi-modal pixel circuits for biomedical imaging [87]. It operates in active readout mode for low-intensity sensor signals such as fluoroscopy in X-ray imaging, and in passive

46

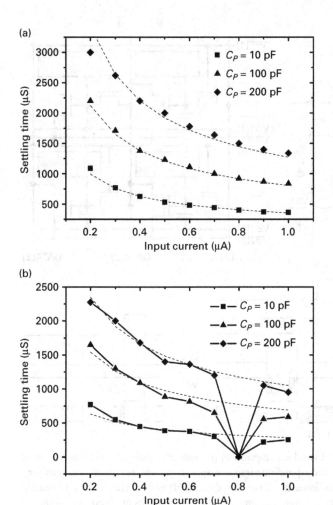

Figure 3.1 Settling time of current programming as a function of initial voltage: (a) $V_0 = 0$ V and (b) $V_0 = 4$ V [87].

readout mode for high-intensity sensor signals such as digital radiography. In these circuits, T1 is the amplifier and T2 is the reset switch transistor. But, T1 also performs the function of a readout switch for the active readout. The corresponding array structure is depicted in Figure 3.2(b). Figure 3.2(c) describes the active mode operation of the

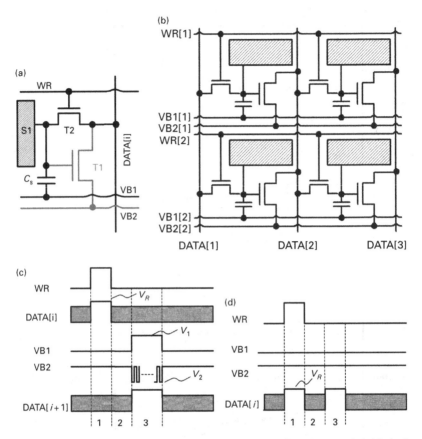

Figure 3.2 2-TFT hybrid active-passive pixel sensor circuits: (a) 2-TFT hybrid pixel circuit and (b) its corresponding array structure along with (c) signal diagrams for active readout (low-intensity sensor signals) and (d) passive readout (high-intensity sensor signals) mode signal diagram. The gray components are used only in the active readout mode.

pixel circuits. Initially, during the reset cycle, the gate voltage of T1 is charged to a reset current (I_R). During the programming, I_R is larger than the required current to accelerate the settling time. Figure 3.3 shows the settling time of the pixel circuit using 4 μA where $V_T[n]$ and $V_T[n-1]$ represent the threshold voltage of the two adjacent pixel circuits in the array. It is clear that the settling time is below 20 μs

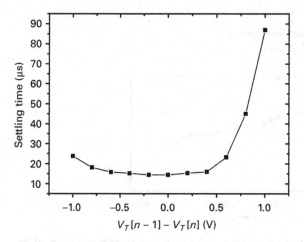

Figure 3.3 Settling time of CBVP pixel circuit [87].

for a wide range of threshold voltage variation. This is the case for sensor application since the pixel ages are almost identical. Also, since the short-term stress condition introduced by Chaji [88, 89] is adopted, the aging of the pixel is limited significantly.

During the integration cycle, where T1 is not under stress ($V_{GS} = V_{DS} = 0$), the sensor signal is integrated into the storage capacitor and generates a voltage (V_{gen}), which modifies the reset current (I_R) and changes the output current accordingly. The change, associated with the output current, is given by

$$\Delta I = g_m V_{gen}, \quad g_m = 2\sqrt{KI_R}, \quad V_{gen} = \frac{Q}{C_S}, \quad \text{and} \quad Q = \eta t_{int} \quad (3.1)$$

where K is the gain parameter in the I–V characteristics of the TFT [65], Q the charge generated by the sensor, t_{int} integration time, and η the conversion rate which is a function of the conversion efficiency of the sensor, input signal intensity, and sensor area. The current of T1 can be read out through the same data line (DATA[i]) by a trans-resistance or charge amplifier while VB1 and VB2 are at V_1 and V_2 voltages, respectively. These voltages should be chosen in such a way

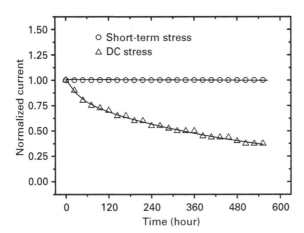

Figure 3.4 Stability of the pixel current (here, voltage programming is used for DC biasing).

that T1 is turned on during the readout cycle. For example, VB1 can be connected to ground during all the operating cycles ($V_1 = 0$). Also, VB2 has the reset voltage (V_R) during the reset and integration cycles and zero ($V_2 = 0$) during the readout cycle. Therefore, T1 is ON only during the readout cycle, minimizing the effect of the leakage current on the operation of the pixel circuits in the same column. More significantly, since T1 is ON for only a fraction of the frame time, it remains stable for a longer time [89]. Figure 3.4 signifies the pixel current stability for an over 500-hour operation.

During the readout cycle, a periodic pulse is applied to either VB1 or VB2, turning T1 on and off periodically. Since the flicker noise at low frequencies stems from the large time constant trapping–detrapping events, the applied pulse, depending on the pulse width, amplitude, and frequency, serves to reduce the noise at lower frequencies similar to what has been observed in the transistors in CMOS technology [90, 91]. Considering the fact that flicker noise is one of the most important limiting factors of the performance in readout circuits, the switched biasing technique can improve the noise performance of the pixel

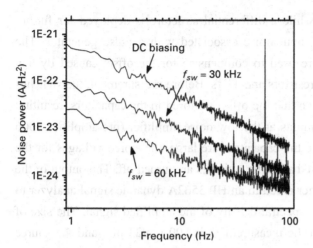

Figure 3.5 Pixel output current noise using different biasing techniques.

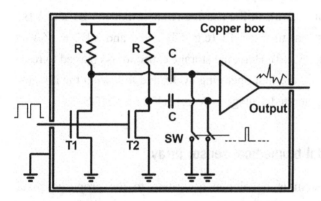

Figure 3.6 Measurement setup for TFT $1/f$ noise.

significantly. The noise performance of the pixel is depicted in Figure 3.5. Although, the noise during a normal readout (30 kHz readout) is smaller than DC biasing, it is larger than the switched biasing technique with two pulses per readout cycle (60 kHz). The setup used for measurement of the flicker noise is shown in Figure 3.6. Here an 8013B HP pulse generator is used for the biasing signal. The entire setup is placed in a double-shielded copper box to reduce external noise coupling, while a differential structure is deployed for further reduction in the input noise associated with a pulse generator. The switches (SW) are used to compensate for the offset caused by any mismatch in the resistors and TFTs. Before measurement, the switches are turned on, such that the offset is stored in the capacitors, resulting in the balanced inputs at the low-noise amplifier (pre-amplifier 5006 Brookdeal). Since the capacitors are large, the stored voltages for the offset are preserved after the switches are turned off. The output of the low-noise amplifier is fed to an HP 3562A dynamic signal analyzer to extract the power spectral density of the amplified signal. The size of the TFTs used in the measurement is 800 μm/23 μm, and the source and drain terminals are connected to 0 V and 12.5 V, respectively.

The configuration of the pixel circuits in passive readout is exhibited as black in Figure 3.2(a). In the passive readout mode, VB1 and VB2 remain at levels that turn off T1 (e.g. VB1 $= 0$ and VB2 $= V_R$) as depicted in Figure 3.2(d). Here, the storage capacitor is charged to reset voltage (V_R) during the first operating cycle, and following the integration cycle, the generated voltage is read back through T2.

3.2 Multi-modal biomedical sensor array

A complete system is developed according to the proposed pixel circuit. Also, the sensor is a biomedical sensor in which a sensing pad is surrounded by four reference pads [92]. Figure 3.7 shows the block diagram of the micro-array biosensor, including the sensing area and peripheral circuitry. The readout and biasing circuitries are shared among the columns and the pixel circuits provide for a first stage gain and access to the sensing pads. After the pixels are biased with the initial current, the modulated current is read back by the operational trans-resistance amplifier (OTRA). For more details on the design of separate basic analog blocks one can refer to [107].

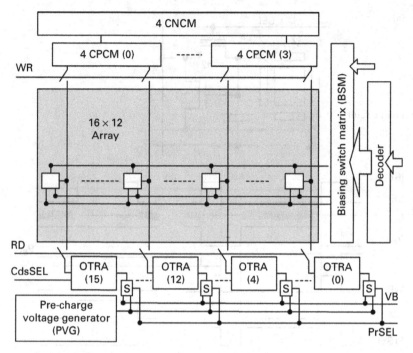

Figure 3.7 Block diagram of the micro-array biosensor.

The pixel circuit used in this array is represented in Figure 3.8 along with the corresponding signal diagram for different modes of detection. The pixel circuit is connected to a sensing pad which is surrounded by reference pads. For the detection modes associated with faint signals such as amperometric or voltammetry [92, 93], integration over time is used to amplify the input signal (see Figure 3.8(b)). The pixel is biased with a constant current during the reset cycle. To accelerate the biasing, the line is pre-charged, and a larger current is employed to bias the circuit. By bootstrapping, the pixel current is dropped to the required level at the end of the reset cycle [87]. During the readout cycle, the current of n1 is read by the corresponding OTRA, connected to the column line. Figure 3.8(c) signifies the operation of the pixel circuit for detection modes with

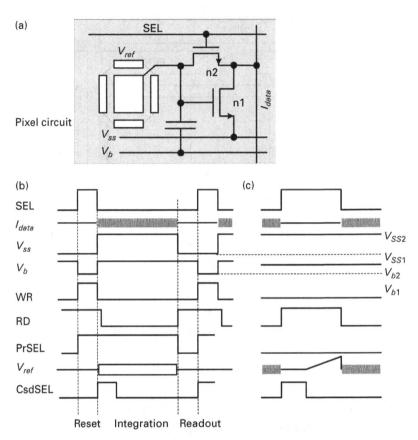

Figure 3.8 Pixel circuit and corresponding signal diagram for different detection modes (* for some detection modes such as voltammetry a time-variant voltage should be applied to V_{ref}).

larger signals such as impedance spectroscopy (IS) [92]. Here, n1 is OFF and the OTRA is directly connected to the sensing pads for reading the current caused by the time-variant voltage applied to the reference pads. The OTRA is reset by a reference current that can be a sensing pad which is not covered by any bio-acceptor material. Thus, the effect of the current caused by electrolyte is subtracted from the final output signals. Although the pixel circuit has been implemented in

CMOS for easier integration, the pixel circuit can be easily replaced with the TFT-based pixel circuit demonstrated in Figure 3.11.

3.2.1 Peripheral circuitries

The biasing switch matrix (BSM) block, switching the biasing configuration of the pixels, is shown in Figure 3.9. When a row is selected for the reset cycle, $V_{b\text{-}en}$ and $V_{ss\text{-}en}$ are high such that V_b and V_{ss} of the pixels in that row are connected to V_{b1} and V_{ss1}, respectively. Here, the pixels are reset with larger current to accelerate the settling. After the reset cycle terminated, the V_b of the pixels are switched back to V_{b2}. As a result, the pixels' current is reduced. If a row is selected for the read cycle, the $V_{ss\text{-}en}$ is high so that the V_{ss} of the pixels is connected to V_{ss1}, allowing the modulated current to pass through n1.

To bias the pixel circuits, a two-level calibrated current mirror is adopted. The first level consists of four NMOS current mirrors (CNCMs) which are calibrated, in turn, with a single input current as observed in Figure 3.10(a). To stabilize the output current of the calibrated current sources over time, a feedback loop is used to set the V_{DS} of p7 to zero. As a result, the leakage current of p7 drops significantly preserving the stored voltage in C3 for a longer time. Each of the CNCMs is used to calibrate four PMOS current mirrors

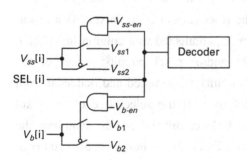

Figure 3.9 Biasing switch matrix (BSM) circuit diagram.

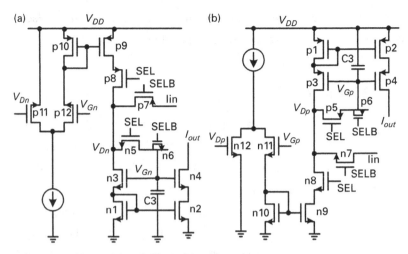

Figure 3.10 (a) NMOS and (b) PMOS calibrated current sources.

(CPCMs) which are connected to a column line through a write switch. As shown in Figure 3.10(b), the same feedback mechanism is used to reduce the effect of leakage current in CPCMs.

An OTRA is used at each column to read out the modulated current of the pixel circuit and also to assist the current sources with pre-charging the column line during the reset cycle, apparent in Figure 3.11. To improve the offset, mismatch, and noise performance of the micro-array, a sample and hold circuit is used for the OTRA biasing (correlated double sampling (CDS) circuit). Before reading out the pixel signal, the CDS circuit samples a reference current which is the reset current of the pixel. As a result, the leakage current of the pixels, connected to a column line (I_{data}); mismatch current of CPCM and/or pixel circuits; offset of the OTRA; and $1/f$ noise of the circuits are sampled and deducted from the final readout signal. Moreover, if the select line of the pixel circuit goes to zero before the CDS circuit finishes the sampling, the charge injection effect associated with the switching is also compensated (see Chapter 6 for more details).

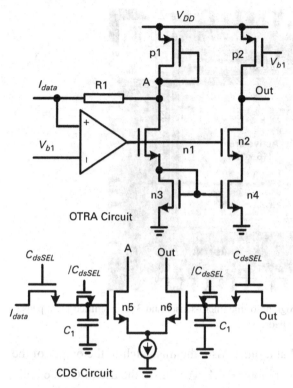

Figure 3.11 OTRA circuit diagram.

3.2.2 Measurement results

Figure 3.12 is the photomicrograph of the fabricated chip in 0.35-µm TSMC CMOS technology. The electrical performance of the array is listed in Table 3.1. Because several test pads are used for verifications the size of the chip is larger than the actual size of the array.

Since the readout circuits used here are simple and shared among the columns, the total power consumption is substantially reduced which allows the up-scaling of the array for higher throughput without the risk of increasing the surface temperature. Also, the power consumption associated with the sensing area is very low (0.4 mW), since only

Figure 3.12 Photomicrograph of the micro-sensor and 3×2 magnified pixel pads surrounded by reference pads.

one row is activated at a time. Also, the mismatch in the output of the OTRAs is exceptionally low (< 24 mV) due to the use of CDS circuit.

Figure 3.13 highlights the stability of the calibrated current sources over time. During the calibration cycle, the current settles to its final value. After n7 turns off, the source driver contains its current for the driving cycle. The drop in the pixel current, due to the leakage, is less than 0.3%. Also, since a dummy transistor and large storage capacitor (C3 = 350 fF) is used to compensate for the charge injection effect of n5 (Figure 3.10 (a)), the change in the output current is negligible.

Since the signal in the amplification mode is DC, having a low noise power at the vicinity of DC is crucial. Thus, both the CDS and switching bias (SB) techniques are utilized in this micro-array for $1/f$ noise reduction. To implement the SB technique, a pulse is applied to the source of n1 during the readout cycle. The noise measurement

Table 3.1 *Electrical performance of the fabricated chip.*

Chip	Values
Technology	0.35 μm (4 metal) at 3.3 V
Power consumption	15 mW
Die size	3.8 mm × 3.8 mm (it is pad limited due to the use of several test pads)
External components	OTRA resistor (R1)

Array	Values
Size	16 × 12

Area	1.71 mm × 1.65 mm
Pixel pitch	140 μm × 100 μm (size depends on the sensor technology)
Operation mode	Impedance spectroscopy, amperometric, cyclic voltammetry
Power Consumption	0.4 mW

Peripheral circuitry	Values
OTRA	Power (0.7 mW), area (102 μm × 107 μm), gain (0.2 MΩ), output mismatch ($<$ 24 mV), 3 db bandwidth (7.5 MHz), and output dynamic range (2 V)
Calibrated current sources	Power (1.5 mW for 16 CPCMs 4 CNCMs) and area (58 μm × 54 μm), mismatch (21 nA at 4.5 μA)

setup is depicted in Figure 3.14 in which a differential amplifier is used to reduce the correlated noises induced by cross talk (such as 60 Hz). Figure 3.15 emphasizes the noise performance of the circuit using both the SB and CDS techniques. Due to the existence of a small mismatch ($<$24 mV) among the OTRAs, some pulses appear at the harmonics of the CDS frequency. The higher the CDS frequency is, the lower the $1/f$ noise is. However, this would reduce the integration time thus limiting

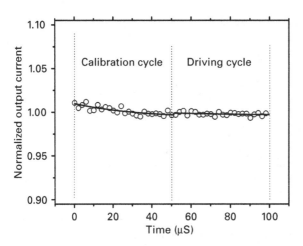

Figure 3.13 Current stability measurement for the calibrated current sources.

the frequency of the CDS. The noise power at the vicinity of the DC is reduced to −83 dB by means of the CDS and SB.

The impedance characteristics of the sensing pads change in the presence of an analyte [92]. Such changes can be measured by applying a sinusoidal signal to the reference pads and tracing the change in the current between the reference and sensor pads. Measurement results for the two different concentrations of bovine serum albumin (BSA) are depicted in Figure 3.16. The result indicates that the impedance consists of a significant capacitance, which changes with the concentration of the BSA. However, the pixel circuit and data line form a low pass filter which reduces the input signal as the stimulus frequency increases which, in turn, saturates the output of the OTRA at high stimulus frequencies. According to Figure 3.16, the chosen frequency must be less than 60 kHz.

For a low intensity of BSA, however, the pixel is configured in the active readout mode signified in Figure 3.17. A ramp voltage is applied to the reference pads, and the current is integrated in C_S. Here, the pixel remains connected to the OTRA to display the transient of the active readout measurement. As described, the CDS circuit can reduce the

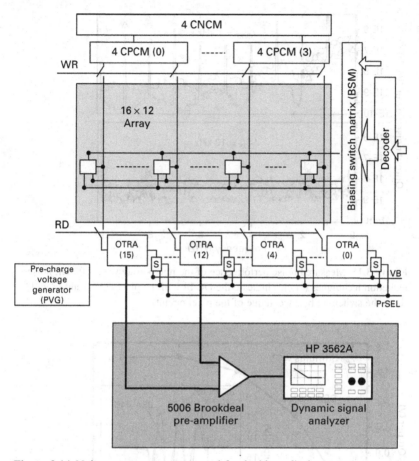

Figure 3.14 Noise measurement setup used for the biomedical sensor array.

leakage and charge injection effects of the pixels. During the reset cycle of the gray curve, the select of the pixel circuit is turned off before the CDS circuit finishes the sampling. As a result, the output voltage of the OTRA settles back to its original voltage (see Figure 3.17). However, for the black curves, the select lines of the pixel and CDS circuits are turned off at the same time. Thus, the OTRA output jumps up at the beginning of the integration cycle (the SB technique is turned off in this measurement for more clarity).

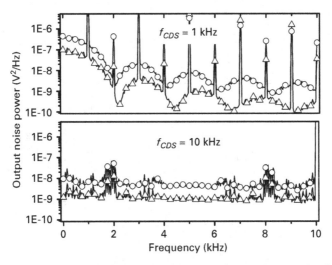

Figure 3.15 Measured noise performance of the micro-array with different noise reduction techniques (circles indicate the DC bias of the pixel circuit and triangles show the switching bias technique of the pixel circuit).

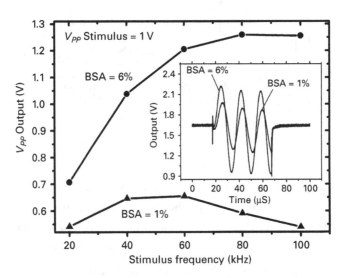

Figure 3.16 Measurement results for impedance spectroscopy.

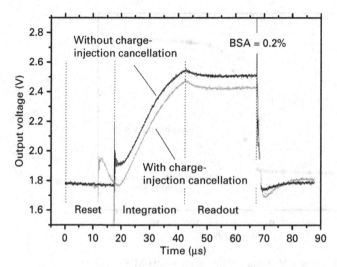

Figure 3.17 Measurement results for the effect of charge injection on the output of OTRA.

3.2.3 Improved dynamic range

As explained in (3.1), the generated voltage is a reverse function of the storage capacitor. Consequently, it is desirable for the low-intensity signal to have a small storage capacitor to improve the signal-to-noise ratio (SNR) [94]. On the other hand, for a high-intensity signal, the capacitor should be large to avoid saturating the readout circuitry.

Since the pixel, depicted in Figure 3.2, provides access to the capacitor, a metal–insulator–semiconductor (MIS) structure is employed to change the capacitor value for the different measurement modes. The capacitance–voltage (CV) characteristics of a MIS structure are demonstrated in Figure 3.18. As highlighted, the MIS capacitor can accommodate the need for variable capacitor in a multi-modal imaging/sensor structure. Moreover, the MIS capacitor can provide for gain-boosting for detecting extremely low intensity signal such as DNA/photon count application. Figure 3.19(b) represents the simulation results for the gain of the proposed pixel circuit with and without MIS capacitor gain. To employ the MIS gain, the V_T of the

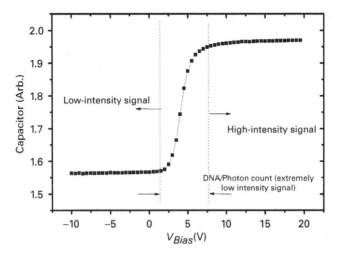

Figure 3.18 Measured CV characteristics of an a-Si:H MIS structure.

MIS should be extracted carefully, to bias the MIS capacitor in an adequate point. Chapter 6 describes a method for extracting the V_T of drive/amplifier TFT. The same technique can be used to detect the edge of the MIS capacitor.

3.2.4 Noise analysis of CBVP pixel circuit

Figure 3.20 shows the small circuit model used to analyze the input referred noise caused by the pixel components. During the reset cycle, node B is assumed to be float since it is connected to a current source. The reset noise associated with T1 and T2 is calculated as

$$V_{Rn1} = \frac{i_{n1}}{(C_T s + g_m)} \text{ and } V_{Rn2} = \frac{R_2}{r_d} \frac{i_{n2}}{\left((1 + \frac{R_2}{r_d})C_T s + (g_m + \frac{1}{r_d}) \right)}$$

$$(3.2)$$

where, r_d is the drain-source resistance of T1, and R_2 the switch resistance of T2. Assuming that $r_d/R_2 \ll 1$, the effect of T2 is attenuated significantly.

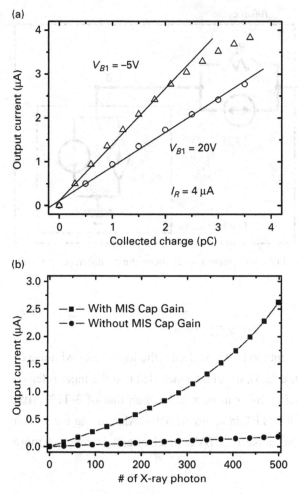

Figure 3.19 (a) Active mode gain adjusting with MIS capacitor and (b) pixel sensitivity with and without MIS capacitor gain.

During the readout cycle, we assume that node B is connected to a resistive load. Also, the effect of C_{gd} of T1 becomes important during the readout cycle. The input referred noise during the readout cycle can be calculated as

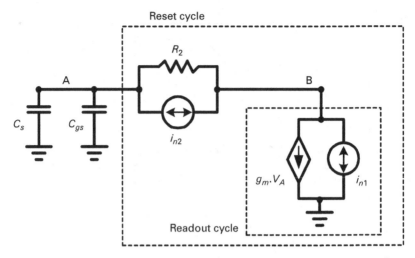

Figure 3.20 Noise model of CBVP imager pixel circuit during different operating cycles.

$$V_{Rdn1} = \frac{i_{n1}}{\left(\frac{C_T + C_{gd}}{R_L C_{gd}} + g_m + C_T s\right)}. \tag{3.3}$$

Here, the R_L is the output load. Considering the input referred noise of 3-TFT APS, (3.2), and (3.3), it can be concluded that the input referred noise of CBVP pixel circuit can be smaller than that of 3-TFT APS, since there is no switch TFT in series with T1. Also, using the switch biasing technique the noise of CBVP pixel circuit is reduced more significantly.

3.3 CBVP AMOLED pixel circuit

For a technology, such as poly-Si, in which mobility experiences variations as well as threshold voltage mismatches [32], a current programming is prerequisite for high yield. Since the CBVP benefits from advantages of both current and voltage programming, the CBVP pixel circuit is a suitable candidate for the above mentioned technology. Figure 3.21 offers the proposed CBVP pixel circuits [95], providing

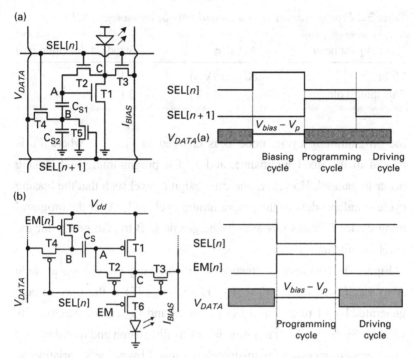

Figure 3.21 CBVP AMOLED pixel circuits along with the corresponding signal diagram: (a) n-type TFT pixel circuit [87] (b) p-type TFT pixel circuit.

for large-area high-resolution AMOLED displays. Here a fixed large bias current (I_{bias}) is used to compensate for the aging and mismatches. Since the current levels required by OLED are around 1 μA, the bias current is divided inside the pixel by a bootstrapping technique. During the biasing cycle, SEL[n] which is the address line of the nth row is high and the voltage at node A adjusts to $V_{bias}+V_T$ where $V_{bias}=(I_{bias}/K)^{0.5}$. The OLED needs to be OFF during the biasing cycle, and so $V_{DD}-V_{bias}-V_T$ should be smaller than the ON voltage of the OLED in Figure 3.21(a). However, in the pixel circuit depicted in Figure 3.21(b), emission TFT turns off and so the OLED is disconnected during the programming cycle. Since I_{bias} is fixed, the line voltage is approximately constant resulting in a significantly fast settling time. During

Table 3.2 *Process variation in a typical poly-Si technology [32].*

Process parameter	Average	Standard deviation (sd)
Mobility	80 (cm^2/V s)	3.3
Threshold voltage	1 (V)	0.1

the programming cycle, node B is charged to $V_{bias} - V_P$ where V_P is related to the pixel luminance, and so the programming and biasing occur in parallel. However, one can design a pixel such that the biasing cycle is independent of the programming cycle [87]. After the programming cycle, T5 turns ON and discharges node B to zero or V_{dd} for the pixel circuits in.

Figure 3.21(a) and (b), respectively, changing the voltage at node A to $V_{bias}(T1) - V_{bias} + V_P + |V_T|$. Here, $V_{bias}(T1)$ is the bias voltage generated by T1 related to the bias current and V_{bias} is the generic bias voltage. Since $V_{bias}(T1)$ is a function of its dimension and mobility, this pixel can compensate for mismatches caused by process variation as well. During the driving cycle, the current of T1 which is controlled by its gate voltage drives the OLED independent of the V_T of T1.

To verify the tolerance of the pixel circuit, a Gaussian distribution for the mobility and V_T variation typical for poly-Si (depicted in Table 3.2) [32] and Monte Carlo simulations are used. Simulation is carried out for 240 pixels resembling a row in a QVGA display (240×3×320). Also, the simulations are conducted for a pure voltage-programmed pixel circuit [96]. Figure 3.22 demonstrates the simulation results as a visual artifact on the display quality. The simple 2-TFT pixel suffers from both V_T and mobility variation and the voltage-programmed pixel circuit cannot handle mobility variation, whereas the CBVP compensates for both variations.

Also, to investigate the effectiveness of the new CBVP pixel circuits, the circuits are assembled from a pre-fabricated pixel circuit and a discrete TFT (see Figure 3.23). To test a single pixel, T4 and T5 can

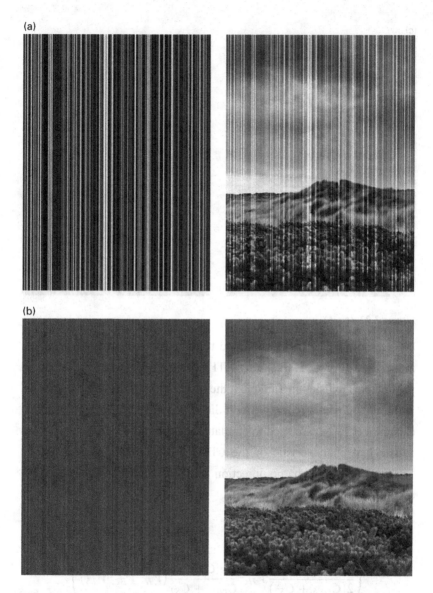

Figure 3.22 Monte Carlo simulation results: (a) conventional 2-TFT, (b) voltage-programmed, and (c) CBVP pixel circuits. Color versions of these figures are available online at www.cambridge.org/chajinathan.

(c)

Figure 3.22 (*cont.*)

be the same TFT. Thus, to reduce the effect of parasitic capacitance induced by discrete TFTs, a single TFT is used for T4 and T5. The pixel parameters are listed in Table 3.3, the aspect ratio of the discrete TFT is 300 μm/23 μm, and I_{bias} is 4 μA. A diode-connected TFT (T_{OLED}) and a capacitor (C_{OLED}) are used to emulate the OLED.

The I–V characteristic of the CBVP pixel circuits is depicted in Figure 3.24. The charge injection and clock feed-through can be modeled as follows

$$V_5 = \frac{C_{S1}}{2C_{OV1} + C_{OV2} + C_{S1}} \left(\frac{C_{OV5}}{2(C_{OV5} + C_{S2})} V_H - \frac{C_{gs5}}{C_{OV5} + C_{S2}} (V_H - V_{B2} - V_{T5}) \right). \tag{3.4}$$

Lifetime measurement result of the AMOLED CBVP pixel circuits is shown in Figure 3.25. Here, the source voltage of T1 is increased to emulate the V_T-shift. The increase in the current is due to the charge

Table 3.3 *AMOLED CBVP pixel circuit parameters.*

Name	Description	Values
W/L(T1)	Aspect ratio of T1	400/23
W/L(T2)	Aspect ratio of the TFT used for S1	100/23
W/L(T3)	Aspect ratio of the TFT used for S2	100/23
W/L(T4)	Aspect ratio of the TFT used for S3	100/23
C_S	Storage capacitance	1 pF
V_H	ON voltage of the switches	30 V
V_{DD}	Operating voltage	20 V

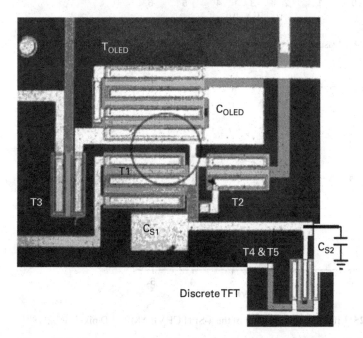

Figure 3.23 Photomicrograph of the fabricated AMOLED CBVP pixel circuit [87].

injection as discussed in Chapter 6. However, drop in the current is due to the channel length modulation [65] since the drain-source voltage (V_{DS}) is decreased by the artificial aging. This will not be the case in a real aging experience since V_{DS} is fixed.

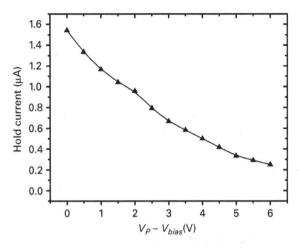

Figure 3.24 Measured I–V characteristics of the pixel circuit shown in Figure 3.21(a) [87].

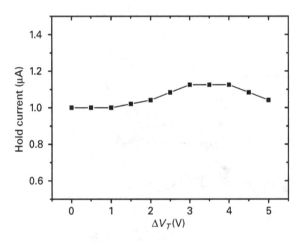

Figure 3.25 Lifetime measurement of the a-Si:H CBVP AMOLED pixel circuit [87].

3.4 Summary

The CBVP driving scheme can improve the settling time and benefits from the simplicity of the voltage programming and accuracy of current programming. The CBVP driving scheme can compensate for all process variations including mobility and V_T mismatches leading to higher yield especially in poly-Si. The driving scheme is easily adapted in a high-resolution multi-modal biomedical imager. Here, in-pixel gain is used for low-intensity signals whereas high-intensity signals are detected in the passive mode. Also, the pixel accommodates a variable capacitor by using a MIS structure for further dynamic range improvement. For low-intensity signals, the capacitor is biased to provide a lower capacitance resulting in a higher SNR. For high-intensity signals, the capacitor is biased for higher capacitance value preventing saturation of the external driver.

The CBVP driving scheme is also used in the AMOLED pixel circuit. Here, the settling time fits most applications but for larger displays, there is a need for faster settling driving schemes. These are discussed in the next chapter.

Copyright notices

4 Enhanced-settling current programming

Although the current mode active matrix provides an intrinsic immunity to mismatches and differential aging, the long settling time at low current levels and large parasitic capacitance is a lingering issue. As explained in Chapter 2, the major source of the settling time in current programming is the large parasitic capacitance. Although using small switch transistors can reduce the parasitic capacitance to some extent [17], the optimized settling time still cannot fit the programming time required for most applications. In addition, the acceleration techniques cannot improve the settling time significantly, particularly if low mobility technologies are used. As a result, driving schemes are required that can reduce the effect of parasitic capacitance based on localized current sources or external driver assistance. This chapter reviews the local current source, current feedback mechanism, and positive feedback described in [98] and [99].

4.1 Localized current source

This technique is more useful for compensating for a threshold voltage shift under bias stress and does not work well for process variation. However, it can be used in conjunction with external calibration techniques described in Chapter 6. The idea is to develop the current sources for each pixel locally and therefore eliminate the parasitic capacitance altogether. Our studies show that most TFTs including a-Si:H TFTs are stable if they are under bias stress for a fraction of frame time. Figure 4.1 shows the results of stressing a TFT under

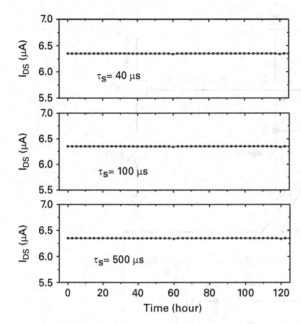

Figure 4.1 Measurement data of TFT stability under short-term bias stress [88].

6.5 μA for different stress time during 16 ms frame time. As it can be seen, the output current of the TFT is stable over time even at 500 μs stress time. A local current source based on the short-term stability of the TFTs is demonstrated in Figure 4.2. Here, T_{LCS} converts the voltage stored on its gate to a current. The current in turn adjusts the source voltage of the drive TFT to allow the entire current to pass through the drive TFT. After the programming is finished, the gate voltage of T_{LCS} turns to a reset voltage and so T_{LCS} stays OFF for the rest of the frame time. Figure 4.3 demonstrates the compensation performance of the local current source driving scheme. Here, the threshold voltage of T1 is modified in the simulations and its effect is captured on voltage created at node B and the output current of T1. The results display that the current remains stable despite 3-V shift in the V_T of T1. At the same time, the voltage created at node B changes in line with shift in the V_T to compensate for it.

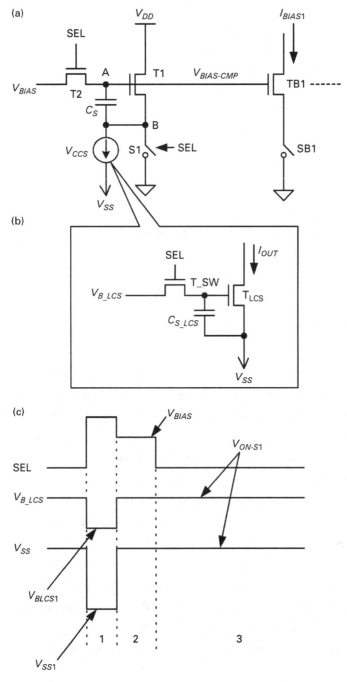

Figure 4.2 Enhanced current programming based on local current source [89]: (a) biasing circuits, (b) local current source, and (c) timing diagram.

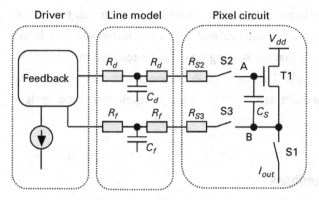

Figure 4.3 Enhanced current programming based on current feedback adapted from [17].

4.2 Current feedback

Figure 4.3 highlights the basics of the current-feedback driving scheme. A real-time feedback loop is used to monitor the output current of the driver TFT and adjust the gate voltage accordingly. As can be seen, the driving scheme is dealing with parasitic components in two lines: data and feedback which can easily result in undesired settling behavior such as overshoot and oscillation. One method to reduce the effect of parasitic in the feedback line is the use of an operational trans-resistance amplifier (OTRA) [17]. In this method, the feedback line is virtual ground, and so its effect is reduced dramatically. At the same time the difference between the programming current and the current running through the feedback line is being translated to voltage and applied to the gate of T1. The analysis of this driving scheme presented in [17] shows that g_m of T1, switch resistance ($R_{S2} + R_{S3}$), and C_S have the major effect on the settling behavior of the circuit. The dependency on g_m of T1 results in different settling time for different programming current due to the large range of the programming current required for AMOLED displays for example. A proposed method in [17] is the use of pulse current as the programming current consists of a larger current at the beginning and actual required current after a few micro-seconds. It has been demonstrated that the settling time can be reduced to 40 μs even for low current levels. This can enable the use of current programming for small to medium size devices with programming time budget around 50 μs.

4.3 Positive feedback

As discussed in Chapter 2, large parasitic capacitance along with small g_m of the drive TFT causes the large settling in the current programming scheme. Another way to overcome this problem is to generate a negative load to cancel out the effect of such parasitic load on the signal line. Figure 4.4 shows a method to develop such a negative load based on a positive feedback structure [98, 99]. Here, a band pass (BP) filter is used as the feedback function. Although more of the high frequency component of the line transient voltage forces more current through the feedback loop, the high frequency noises will be amplified due to the positive feedback too. Therefore, there is a trade-off between noise and settling time in such a system which will be discussed later.

Since each driver needs to support a few hundred output channels, the size of the output buffer should be minimal to avoid an oversized

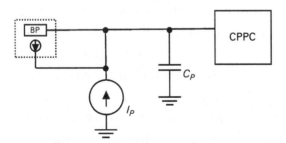

Figure 4.4 Fast current driver based on positive feedback [98].

driver. As a result, the feedback is implemented as a differentiator with one-pole filter for the second cut-off frequency. To implement this structure, the line voltage needs to be copied and pass through a capacitor for creating the current. After that the current needs to be added to the programming current. A circuit block that offers all these functionalities is a simple current conveyer type II (CCII) [100, 101]. The I–V characteristics of a typical CCII are as follows

$$
\begin{bmatrix} I_Y \\ V_X \\ I_Z \end{bmatrix} = \begin{bmatrix} 0 & 0 & 0 \\ A & 0 & 0 \\ 0 & M & 0 \end{bmatrix} \begin{bmatrix} V_Y \\ I_X \\ V_Z \end{bmatrix}. \tag{4.1}
$$

The amplitudes of A and M can be approximated as unity and independent of the input values for small signals. However, since active matrix devices deal with large signals, A and M become input signal dependent. The voltage at the X terminal is controlled by the voltage at Y, which is equal to the voltage at the Z terminal ($V_X = AV_Z$ where $|A| \leq 1$) and the current of X terminal is copied to the Z terminal by a gain of M. Figure 4.5(a) and (b) shows the driver implemented based on CCII. The programming current (I_P) can be either applied to the X or to the Z terminal. Applying the programming current to the Z terminal offers a separate path for the feedback current which provides for more control on reducing the size of the feedback capacitor. Considering the case in which the programming current is directly connected to the Z terminal, the current at the X terminal is given by

$$
I_X = -AC_F \frac{d}{dt} V_Z, \tag{4.2}
$$

where C_F is the feedback capacitor, and V_Z the voltage at the Z terminal. As described in (4.2), the feedback capacitor C_F acts as a negative capacitance which can be used to reduce the parasitic capacitance effect. Since the Z terminal is connected to the pixel during the programming time, the behavior of the system can be written as

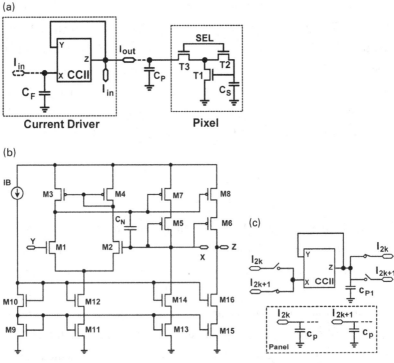

Figure 4.5 (a) Fast current driver, (b) circuit diagram of the current conveyor II, and (c) implementation method for C_F using column parasitic capacitance [99].

$$I_P = (C_P - AMC_F)\frac{d}{dt}V_Z + \beta(V_Z - V_T)^2. \qquad (4.3)$$

Based on the discussion in Chapter 2, the time constant of (4.3) can be extracted as $\tau = 2(C_P - AMC_F)/(\beta.I_P)^{0.5}$. Consequently, choosing AMC_F close to C_P reduces the effect of the parasitic capacitance significantly. To reduce the size of C_F, one can use a large M in the current mirror enabling in-chip implementation of the feedback capacitor. Figure 4.5(c) shows a proposed architecture for using the parasitic capacitance as C_F. C_{P1} is a small integrated compensation capacitance to stabilize the driver and the switches can be shared with an offset cancellation circuit as discussed further in the next sections.

4.3.1 Stability and noise analysis

Similar to any feedback system the stability of this driver needs to be reviewed properly, specially due to the use of positive feedback. Since the system is designed for a large signal, the Lyapunov approach [102] is used for stability analysis. According to the Lyapunov approach, a system is unconditionally stable if you can define an energy function such that its first-order derivative is negative. To investigate the stability based on the Lyapunov approach, the settling point of (4.3) is moved to zero as follows

$$\frac{d}{dt}x = \frac{x}{(C_P - AMC_F)}\left(-\beta x - \sqrt{\beta I_P}\right);$$
$$x = V_Z - \sqrt{\frac{I_P}{\beta}}.$$
(4.4)

By defining the energy function as $E(t) = x^2$, the derivative of E can be written as

$$\frac{d}{dt}E = \frac{2x^2}{(C_P - AMC_F)}\left(-\beta x - \sqrt{\beta I_P}\right).$$
(4.5)

Based on the Lyapunov approach, it can be concluded that the driver is stable if (4.5) is negative which means that AMC_F should be positive and smaller than the overall parasitic capacitance. Although, the Lyapunov approach covers both small and large signals, one can use small signal analysis for stability of the driver in the presence of noise, cross talk, and small current levels. Figure 4.6 shows the root-locus plot for the circuit with two different feedback capacitors. The circuit is unconditionally stable for $C_F = 90$ pF ($< C_P$). On the other hand, to stabilize the circuit with $C_F = 110$ pF ($> C_P$), the close loop gain between V_Y and V_X should be smaller than 1. As predicted, the result is similar to (4.5) which

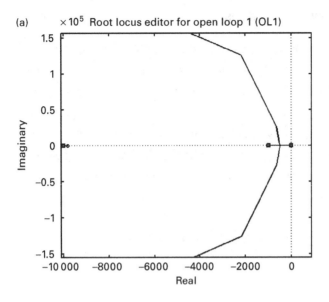

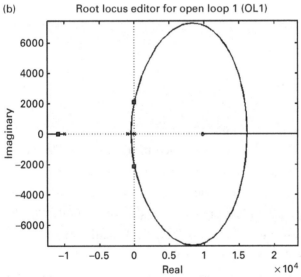

Figure 4.6 Root-locus plot for (a) $C_F = 90$ pF and (b) $C_F = 110$ pF [99].

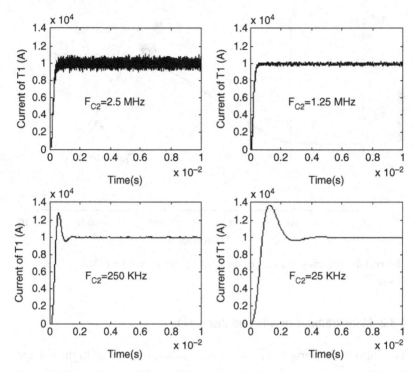

Figure 4.7 Transient waveforms of the positive feedback current source with different cut-off frequencies in the presence of noise [98].

means the driver will be unstable for effective feedback capacitor larger than the parasitic capacitor.

Figure 4.7 emphasizes the effectiveness of the filter on the noise performance. It is evident that increasing the cut-off frequency of the LP filter makes the driver more sensitive to the noise of the current line. However, as the cut-off frequency increases, the speed increases as well. The Butterworth filter is implemented by simply adding a capacitor between the output of the voltage feedback and X terminal (C_{LP}). As is shown in Figure 4.8, increasing the C_{LP} results in a smaller F_{C2}. Also, the circuit acts as a differentiator between F_{C1} and F_{C2}. The slope of the differentiator is defined by the value of C_F.

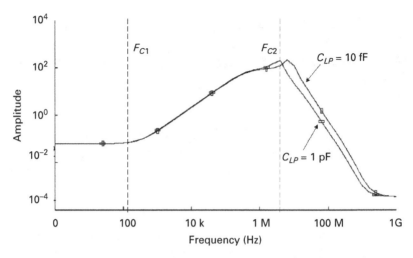

Figure 4.8 Frequency response of the CCII current source for different values of C_{LP} [98].

4.3.2 Measurement results and discussion

The photomicrograph of the driver fabricated in a high voltage CMOS process (the 0.8-μm process of DALSA semiconductor) is given in Figure 4.9. The parameters of the fabricated driver are listed in Table 4.1. Figure 4.10 demonstrates the settling time of the pixel circuit for a step current of 100 nA in the presence of parasitic and feedback capacitances of 100 pF. It is clear that the fast current driver can manage the effect of parasitic capacitance and reduce the settling time significantly.

The I–V characteristics of the current driver are extracted by using a Keithley 236 source measurement unit (SMU). For the results shown in Figure 4.11, the output is connected to a fixed voltage (7 V), while the input current is swept. The output current closely follows the input current at a fixed input voltage of 7 V. However, the voltage dynamic range does not stretch from rail-to-rail (as observed in Figure 4.12). Here, the input current is 1 μA, and the output voltage is swept from

Table 4.1 *Parameters of fabricated current driver.*

Name	Description	Value
$(W/L)_{1,2}$	Aspect ratio of M1 and M2	30 μm /6 μm
$(W/L)_{3,4}$	Aspect ratio of M3 and M4	40 μm /6 μm
$(W/L)_{5,8}$	Aspect ratio of M5 to M8	120 μm /6 μm
$(W/L)_{9,10}$	Aspect ratio of M9 and M10	12 μm /6 μm
$(W/L)_{13,16}$	Aspect ratio of M13 to M16	60 μm /6 μm
$(W/L)_{19,20}$	Aspect ratio of M19 and M20	10 μm /6 μm
C_N	Noise reduction capacitance	1 pF
C_F	Feedback capacitor	10–200 pF
C_P	Parasitic capacitance	10–200 pF

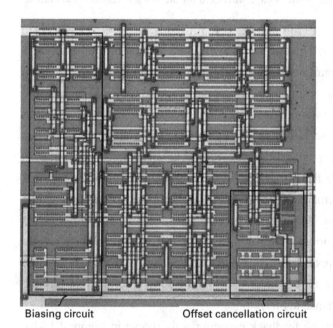

Biasing circuit   Offset cancellation circuit

Figure 4.9 Photomicrograph of the fabricated current driver [99].

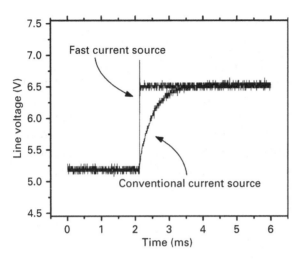

Figure 4.10 Transient line voltage waveform for the fast and conventional current driver [99].

0 to $V_{DD} = 15$ V. This can result in different settling times for different programming currents, as discussed in the following.

Figure 4.13(a) shows the settling time extracted for a 200-pF parasitic and feedback capacitance for the current cell demonstrated in Figure 2.4(a). Here, $V_f - V_0$ is set to 1 V throughout the entire measurement range. Since the programming time for large-area and high-resolution active matrix devices is larger than 10 µs, it is clear that the driver presented here can meet the requirements of a vast range of applications since the settling time is less than 4 µs. While it is expected that the settling time drops as the programming current increases, the measurement results exhibit a reverse trend. This occurs because the closed-loop voltage gain reduces for large voltages, as depicted in Figure 4.12(b), mitigating the effectiveness of the driver in accordance to (4.3).

The effect of the feedback capacitance is shown in Figure 4.13(b). The input current is 1 µA and the parasitic capacitance is 100 pF. The settling time decreases linearly, as the feedback capacitance increases, again corroborating with (4.3).

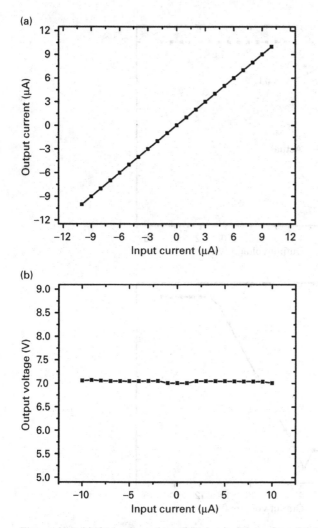

Figure 4.11 (a) Output current and (b) input voltage for a fixed output voltage as a function of input current [99].

4.3.3 Self-calibration of the current source

In applications for which the current driver is intended, the current levels are small (of the order of 1 μA) which compounds the effect of the offsets mandating an effective offset cancellation technique.

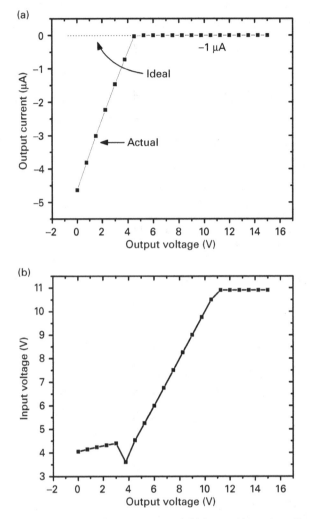

Figure 4.12 (a) Output current and (b) input voltage for a fixed output voltage as a function of input voltage [99].

Figure 4.14 demonstrates two different configurations for the offset cancellation. A differential circuit is adopted in both configurations to reduce the effects of the charge injection and clock feed-through leading to small storage capacitors. In case I (Figure 4.14(a)), the

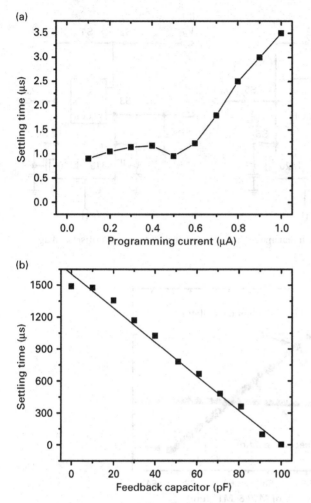

Figure 4.13 Settling time as a function of (a) programming current and (b) feedback capacitor [99].

circuit is disconnected from the input current and output load during the cancellation process. If there is an offset between the current of the X and Z terminals, M19 and M20 store the offset, alleviating its effect [97]. Since the statistics of the process variations for the technology is not available, the offset is emulated by variations in the

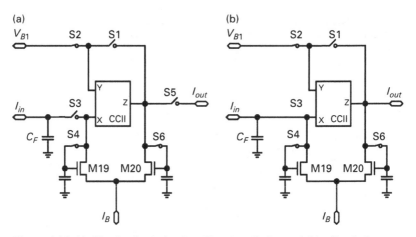

Figure 4.14 (a) Circuit schematics for offset cancellation and (b) offset-leakage cancellation [99].

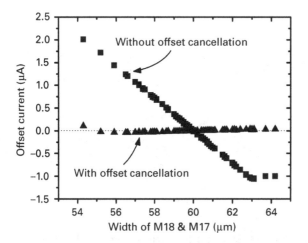

Figure 4.15 Monte Carlo simulation of the offset current as a function of transistor mismatch with and without offset cancellation [99].

width of the paired transistors. Figure 4.15 shows Monte Carlo simulation results for variation in the width of M17 and M18. The offset current varies from -1 µA to 2 µA for the circuit without cancellation. In contrast, the offset current remains smaller than 10 nA for case I. Although this technique can control the offset associated with the

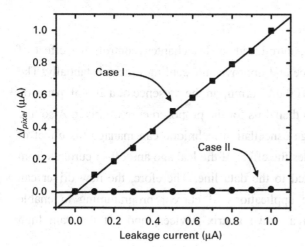

Figure 4.16 Effect of leakage on the output of case I and II of cancellation techniques [99].

driver, the technique is prone to offsets, stemming from the input and leakage currents caused by the pixels connected to the output.

Figure 4.14(b) shows a configuration that compensates for the aforementioned issues in addition to the driver offset. Here, the input and output terminals are connected to the current source and output load, respectively, during the cancellation process. Therefore, any output leakage or source offset is stored by M19 and M20 rendering the circuit offset-immune. Figure 4.16 demonstrates measurement results for which a current source is connected to the output of the driver to emulate the leakage current, and the output of the current cell connected to the driver is monitored. In case I, the leakage current is conveyed to the cell output current, whereas, for case II, the output current is independent of the leakage current. Moreover, this circuit can perform the CDS reducing the reset and programming noises. Since the reset noise is the most significant portion of the input referred noise in sensors and imagers, the noise performance improves significantly. This can be seen in Figure 4.5, where the gain for low frequency signal is around −40 dB.

4.4 Summary

The current driver, introduced in this chapter, controls the effect of parasitic capacitance and improves the settling time substantially. The settling time for a 100 nA current, in the presence of a 200 pF parasitic capacitance, is less than 4 μs for the proposed current driver. Also, the novel offset-leakage cancellation technique can manage the effect of driver offset, besides the effect of the leakage and offset currents from the pixels connected to the data line. Therefore, the new driver can further extend the applications of current programming to enable scaling to large-area active matrix devices, and yet maintain high refresh rates.

Copyright notices

5 Charge-based driving scheme

Considering all the design considerations, an ideal driving scheme for large-area electronics should not only prevent additional complexity in the simple voltage programming (2-TFT pixel circuit [6]) but also compensate for the instability (or mismatches) of the backplane without compromising the aperture ratio. Thus, new driving schemes are needed in which the pixel aging is compensated based on discharged voltage [103, 104]. We developed a scheme along these lines whereby the amount of the leaked charge from the gate voltage of the drive/amplifier transistor changes as the TFT ages and compensates for the aging. A similar technique can be used to adjust the gain of the sensor pixels to provide for a very large programmable dynamic range.

5.1 Charge-based pixel circuit

Figure 5.1(a) demonstrates the proposed simple voltage-programmed pixel circuit. During the programming cycle, node A of the pixel circuit is charged to a programming voltage (V_P). During the next cycle which can be in parallel with the programming cycle of the next row, a part of the voltage, stored at node A, is discharged through Td while S2 and S3 are ON. The amount of discharged voltage, ΔV_A, is controlled by the channel resistance of Td which is defined by the aspect ratio, mobility, and threshold voltage of Td, given by

$$\Delta V_A = \frac{(V_P - V_{OLED} - V_T)^2}{(V_P - V_{OLED} - V_T) + \frac{\tau}{t}} \quad \text{and} \quad \tau = \frac{C_S}{(W/L)_{Td}K}. \quad (5.1)$$

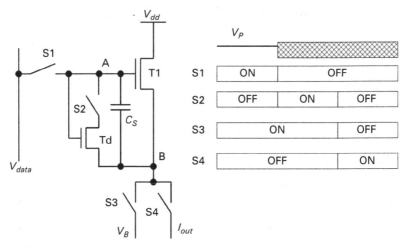

Figure 5.1 Generic charge-based pixel circuit and timing diagram adapted from [104].

Here, V_T is the threshold voltage of Td, V_{OLED} the OLED voltage, $(W/L)_{Td}$ the aspect ratio of Td, K a function of mobility and gate capacitance, and C_S the storage capacitor. More importantly, since Td and T1 are physically adjacent and have the same biasing condition, Td represents T1 in the discharging process. For example, since T1 and Td have the same gate-source bias, the shift in the threshold voltages of T1 and Td due to bias stress is correlated. Also, according to (5.1) a shift in the threshold voltage of Td results in smaller reduction in the voltage at node A for a given discharge time. Therefore, the gate-source voltage of T1 becomes larger, and so compensates for the V_T-shift. Figure 5.2 shows the discharged voltage for a different V_T-shift (ΔV_T). It is clear that as ΔV_T increases the discharged voltage decreases linearly. Here, C_S is 350 fF, T1 is 180 μm/3 μm, T2 is 20 μm/3 μm, T3 is 18 μm/3 μm, and Td is 4.5 μm/4 μm. In a real array, switches are implemented with TFTs as well. In that case the select line of the S2 can be connected to the select line of S1 from the next row. Moreover, S3 and S4 can be replaced with one switch and in an AMOLED pixel circuit, this switch can be the OLED itself.

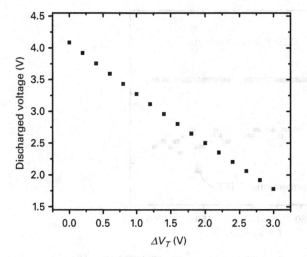

Figure 5.2 The discharged voltage for different shifts in the threshold voltage of T1 [104].

5.1.1 Measurement results

To investigate the effectiveness of the proposed driving scheme, the pixel circuit is fabricated and tested under electrical stress. A microcontroller is used to generate the required signals for a programming time of 20 μs and frame rate of 60 Hz.

Figure 5.3 shows the current passing through T1 during different operating cycles. In this experiment, some delays are added between each operating cycle to exhibit the role of each cycle. During the nth programming cycle, a large current passes through T1 as node A charges to a programming voltage (V_P). The pixel current drops, as T2 turns off at the end of this cycle, due to the charge injection effect of T2. During the ($n+1$)th programming cycle, the voltage at node A starts discharging through Td, and so the pixel current drops to the desired current levels. This current is preserved during the driving cycle since the voltage is stored in the storage capacitor.

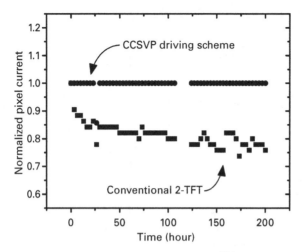

Figure 5.3 Lifetime measurement of AdMo^{TM} and conventional 2-TFT pixels [104].

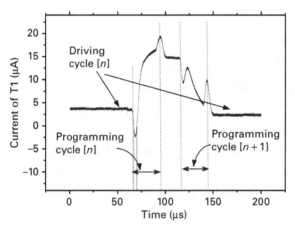

Figure 5.4 Measured current of T1 during different operating cycles [104].

As shown in Figure 5.4, the level of the pixel current (current passing through the OLED during the driving cycle) drops as the timing of the programming cycle increases.

To test the lifetime of the pixel circuit, it is subject to a bias stress which emulates the worst case of operation condition. During each

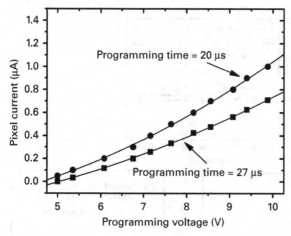

Figure 5.5 Measured I–V characteristics of the AdMo^{TM} pixel circuit [104].

frame time, the pixel is programmed by the maximum programming voltage which results in an initial current of 1.17 μA, and its current is measured. Figure 5.5 compares the lifetime test results of the conventional 2-TFT pixel circuit versus the proposed new driving scheme. The current of the 2-TFT pixel circuit drops by 25% whereas the current of the proposed pixel circuit is significantly more stable. The fluctuation in the current of the conventional 2-TFT pixel circuit is due to the variation in ambient temperature. Since the results for the proposed and 2-TFT pixel circuits have been extracted in the same environment and time, the results indicate that the proposed driving scheme is stable under temperature variations as predicted earlier.

The effect of temperature variation is measured as well. The results are shown in Figure 5.6(a) verifying the temperature stability of the charge based pixel circuit. Here, T_L is the leakage time, and $T_L = 0$ represents the conventional 2-TFT pixel circuit. As suggested in (5.1) increasing the mobility increases the discharged voltage, resulting in smaller gate-source voltage. Thus, the pixel current becomes constant despite the higher mobility. Temperature affects mobility, V_T, and power

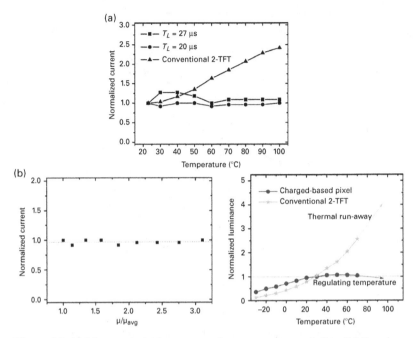

Figure 5.6 (a) Measured pixel current under temperature variation [104] and (b) measured display brightness under temperature variation.

parameter in the I–V-characteristics of the a-Si:H TFTs [104]. Figure 5.6(b) represents the measurement results against the mobility variations (temperature variations are translated to mobility variations using the model presented by C. Ng [105]). This result indicates that the pixel current is stable despite the mobility variations independent of the cause.

5.1.2 Implementation of the relaxation technique

To further extend the lifetime of the display, a new timing scheme is adapted which suppresses the threshold voltage shift. The frame time is divided into programming, compensating, driving, and relaxation cycles [80, 103]. During the relaxation cycle, the gate-source voltage of the

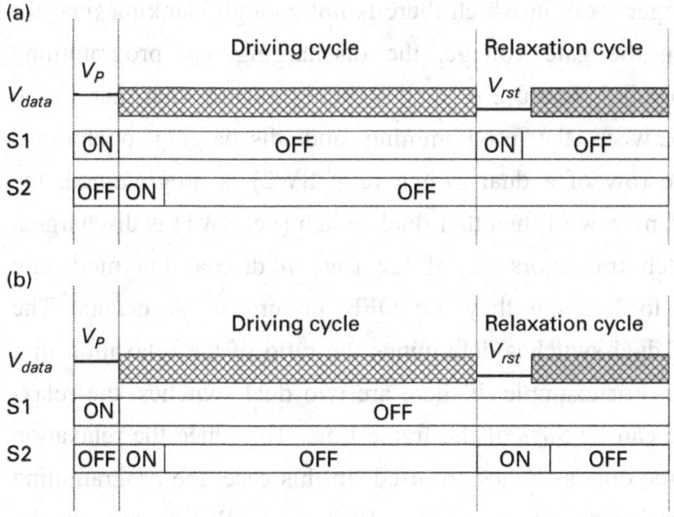

(a)

V_{data}	V_P	Driving cycle		V_{rst} Relaxation cycle	

S1	ON	OFF		ON	OFF

S2	OFF	ON	OFF		

(b)

V_{data}	V_P	Driving cycle		V_{rst} Relaxation cycle	

S1	ON	OFF			

S2	OFF	ON	OFF	ON	OFF

Figure 5.7 Timing diagram for incorporating the relaxation driving scheme: (a) re-programming the pixel with reset voltage and (b) discharging the gate voltage of the drive TFT, adapted from [104].

drive TFT is discharged to zero. Thus, the drive TFT is not under stress, and moreover, the trapped charges are released preventing an accumulation of the aging. There are two ways to implement the new timing: programming the pixel with a reset voltage for the relaxation period, or discharging the gate voltage completely through S2. Figure 5.7 shows the proposed timing diagrams for each method. For re-programming the pixel, S1 turns ON, the value on the V_{data} line changes to the reset voltage (V_{rst}). Thus, node A is charged to the reset voltage which turns the TFT off. The problem with this method is that the programming time needs to be divided by half resulting in shorter row time which can be challenging for larger displays. To eliminate this issue, one can use S2 to discharge the voltage at node A. Here S2 stays ON for a longer time during the relaxation cycle, to make sure the gate voltage is completely discharged. One can use switches on the panel to implement the control signal for S2 from the select signal for S1 [104].

For a larger array in which there is not enough blanking time for discharging the gate voltage, the discharging and programming occur in parallel. The use of dual switches ensures that there is no conflict between the programming and discharging operations. While, one row of a dual switch (e.g. SW2) is programmed, the corresponding row of the other dual switch (i.e. SW1) is discharged. Since switch transistors T2 of the row in discharging mode are connected to VL and they are OFF, no cross talk occurs. The number of dual switches determines the ratio of the relaxation in a frame time. For example, if there are two dual switches, the relaxation cycle can be 50% of the frame time. To reduce the relaxation to 30% three dual switches are used. In this case, the programming and discharging operations occur in SW(i) and SW(i-2) respectively. More importantly, the dual switches can be fabricated with a-Si:H TFTs placed at the edge of the panel.

5.1.3 AMOLED display

The charged basal technique is adapted in the design of 2.2-inch QVGA ($240 \times 3 \times 320$) and 9-inch WXGA ($800 \times 3 \times 480$) displays. Here, each pixel consists of red, green, and blue sub-pixels. To design each sub-pixel, it is necessary to calculate the brightness share of each color for a targeted white point in the tristimulus or CIE coordination [106]. Since each color of the OLED is not pure, meaning it covers a wide range of the spectrum, a 3×3 matrix (T_{SP}) is used to present the color contents of each OLED assuming they light up at 1 nit/m^2. As a result, the brightness share for each sub-pixel is calculated by the following:

$$B = T_{SP}^{-1} W. \tag{5.2}$$

B is a 1×3 matrix containing the brightness share of each sub-pixel, T_{SP}^{-1} the inverse of T_{SP}, and W is the tristimulus of the desired white point. Knowing the brightness share for each sub-pixel and the efficiency of each OLED, the maximum current required for each sub-pixel is expressed as

$$I_i = B_i.\eta_i, \quad i = R, G, B, \tag{5.3}$$

where, I_i, B_i, and η_i are the maximum required current, a value in matrix B, and efficiency, respectively, corresponding to OLED "i." The maximum overdrive voltage of drive TFT (T1) is kept constant and the required aspect ratio of drive TFT is obtained through simulation. This assures that the drive TFT of each sub-pixel ages at the same rate, preserving the white balance point during the display's lifetime. The switch TFTs are designed based on the required settling time.

Moreover, the aperture ratio is adjusted according to the efficiency of each OLED and their brightness share for the white point. As a result the three different OLEDs age at the same rate avoiding any color shift in the white balance. The optimized aperture ratio is given by the following,

$$R_i = \frac{B_i\eta_i}{B_G\eta_G} \quad \text{and } i = R \text{ or } B, \tag{5.4}$$

where, R_i is the ratio of the blue (or red) aperture ratio to the green's aperture ratio. Since green is normally the highest efficiency, green OLED is selected as the reference but any of the other OLEDs can be the reference. The sub-pixel is designed based on (5.3) and during the layout is optimized to achieve the aperture ratios listed in (5.4). The goal is to have the smallest possible overdrive voltage for the TFT and the largest possible aperture ratio for OLED in order to get the best lifetime for the TFT and OLED. Thus, the design of the pixel is an iterative process between selecting an overdrive voltage and obtaining the aperture ratio to get the optimum point for a display lifetime.

Figure 5.8 shows the layout of the RGB pixel for the two display sizes. The maximum front-screen brightness (FSB) for the 2.2-inch display is 200 nit/m^2 (after a 45% efficient polarizer) and it is 150 nit/m^2 for the

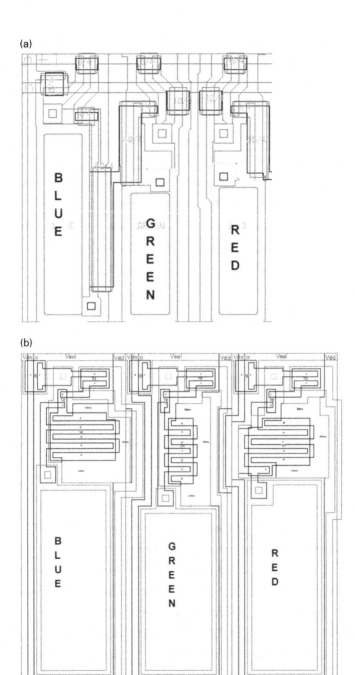

Figure 5.8 RGB pixel layout for (a) 2.2-inch QVGA and (b) 9-inch WXGA displays. Color versions of these figures are available online at www.cambridge.org/chajinathan

9-inch display. The achieved aperture ratios for the 2.2-inch display are 21%, 20.5%, and 30.5% for red, green and blue sub-pixels, respectively. Also, 34.5%, 30%, and 35% aperture ratios are achieved for red, green, and blue sub-pixels of the 9-inch display, respectively.

The fabricated display is measured for the differential aging effect using the gray-scale pattern, depicted in Figure 5.9(a). The image sticking issue is obvious in the conventional 2-TFT display whereas the image is clear for the charged based display.

Also, the brightness of each spot shown in Figure 5.9(a) is measured every day by using a luminance meter (Konica Minolta LS-100). The measured luminance results for the spot numbers 1, 3, and 5 are depicted in Figure 5.10(a). The total error for the maximum stressed part (spot 1) is less than 5% after 200 hours of stress. Since the pixel current is stable, the loss in the brightness can be due to the degradation in OLED brightness. For longer lifetime measurement, a 2.2-inch display was stressed with a typical brightness of 100 cd/m^2 extracted from user pattern. The lifetime depicted in Figure 5.10(b) signifies the stability of the circuit over 6000 hours of continuous operation.

Figure 5.11 provides two pictures of the display after prolonged operation. Since the aging is more rapid at the beginning of the panel life, the results signify the potential of the new driving scheme in fulfilling the lifetime specs for most applications, in particular for portable devices such as cell phones, digital cameras, and DVD players.

5.2 Real-time biomedical imaging pixel circuit

The charge-based compensation driving scheme is used in the design of a real-time imager pixel circuit. Not only does the discharging path compensate for the aging and mismatches, but also it adjusts the gain of the pixel for different applications. The pixel circuit and

(a)

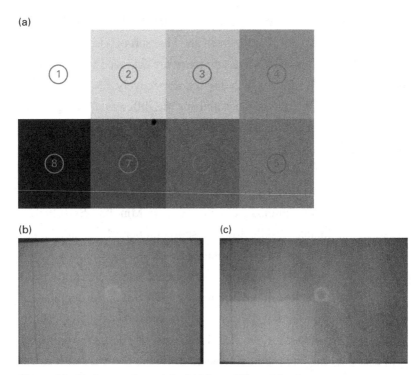

(b)   (c)

Figure 5.9 (a) Gray scale used for 240-hour differential aging measurement along with the measurement result for (b) AdMo^{TM} and (c) conventional 2-TFT. Color versions of figures (b) and (c) are available online at www.cambridge.org/chajinathan.

corresponding signal diagram are demonstrated in Figure 5.12 and the photomicrograph is depicted in Figure 5.13. During the reset cycle, node A is charged to a reset voltage (V_R). The next cycle can be the discharging for compensation. However, since the short-term stress condition is used for the pixel operation, this cycle is ignored. During the integration, the sensor signal is collected by the storage capacitor. During the gain-adjusting cycle, the voltage of the gate leaks out through Td. Leakage time (τ_L) can be adjusted for different applications since it controls the gain of the pixel. During the readout cycle, the pixel current is read through I_{data}. Unlike the other pixel circuits [23] proposed to handle

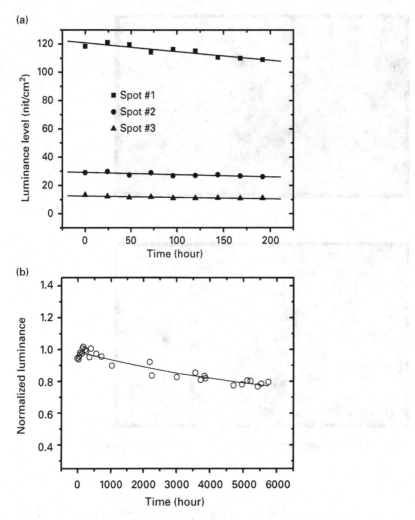

Figure 5.10 (a) Measured luminance stability for different stress levels for 9-inch AMOLED display and (b) the lifetime results of the 2.2-inch display under averaged luminance (100 cd/m²).

high-dynamic-range input signals, the read operation is not destructive, since the proposed pixel circuit operates only in the active mode.

Based on (5.1), the remaining voltage (V_{dmp}) at node A after the gain-adjusting cycle is given by

Figure 5.11 Real images from the 9-inch AdMo™ display [104].

$$V_{dmp} = V_R - V_{gen} - \frac{(V_R - V_{gen} - V_T)^2}{(V_R - V_{gen} - V_T) + \tau/\tau_L}. \qquad (5.5)$$

Here, V_{gen} is the generated voltage due to the collected charge. By assuming that V_{gen} is much smaller than V_R, the linear approximation is employed to calculate the damping effect (A_{dmp}) as the following:

$$A_{dmp} = \frac{1}{1 + \frac{\tau_L(V_R - V_T)}{\tau}}. \qquad (5.6)$$

(a)

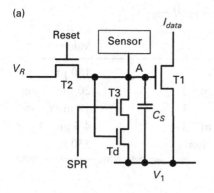

Figure 5.12 Gain-adjustable biomedical imager pixel circuit (a) and signal diagram (b).

(b)

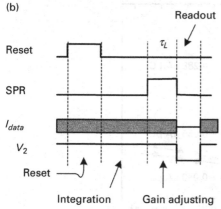

Figure 5.13 Photomicrograph of the gain-adjustable pixel circuit.

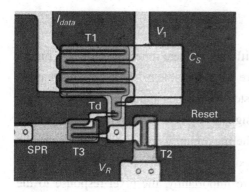

Table 5.1 *Parameters of the fabricated pixel circuit.*

Name	Description	Value
$(W/L)_1$	Aspect ratio of T1	180 μm /3 μm
$(W/L)_2$	Aspect ratio of T2	20 μm /3 μm
$(W/L)_3$	Aspect ratio of T3	18 μm /3 μm
$(W/L)_d$	Aspect ratio of Td	4.5 μm /4 μm
C_S	Storage capacitor	350 fF

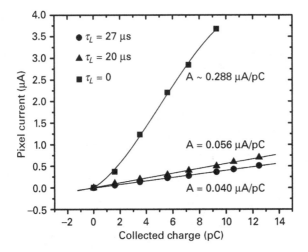

Figure 5.14 Gain-adjustment results using charge-leakage technique.

Measurement results for different leakage times are shown in Figure 5.14. The pixel parameters are listed in Table 5.1. It is obvious that the gain of the pixel can be adjusted for various applications. For example for very low intensity input signals (e.g. fluoroscopy) the leakage time can be close to zero to get the maximum gain. On the other hand the leakage time can be increased (e.g. 27 μs) for higher-intensity input signals (e.g. radiology). More importantly, the pixel response to the collected charge is significantly linear.

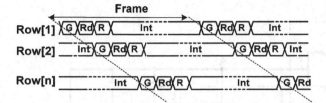

Figure 5.15 Timing schedule for real-time imaging.

The pixel circuit can provide for parallel operation of reset and readout cycles for different rows. As a result, it can be used for real-time imaging applications such as fluoroscopy. Figure 5.15 shows the timing schedule for an array intended for real-time imaging.

5.2.1 Noise analysis of charge-based pixel circuit

To investigate the effect of gain-adjusting branch (T3 and Td) on the noise performance of the pixel, the reset noise of the pixel is calculated with and without the path (the gain-adjusting path does not affect the readout cycle). Figure 5.16 shows the noise model used to evaluate the reset noise of the pixel circuit. The reset noise without considering the gain-adjusting branch is given as

$$V_{n\_2T} = V_{n2} = \frac{i_{n2}}{C_T s + 1/R2} \quad \text{and} \quad C_T = C_s + C_{gs}, \qquad (5.7)$$

in which $R2$ and i_{n2} are the channel resistance and noise of T2, respectively.

If the gain-adjusting branch is used, the noise of T2 is damped. T2 noise effect can be written after damping based on (5.6)

$$V_{n2-dmp} = \frac{i_{n2}}{C_T s + 1/R2} \frac{1}{\left[1 + \frac{\tau_L(V_R - V_T)}{\tau}\right]}. \qquad (5.8)$$

The noise of Td is the dominant noise source in the gain-adjusting

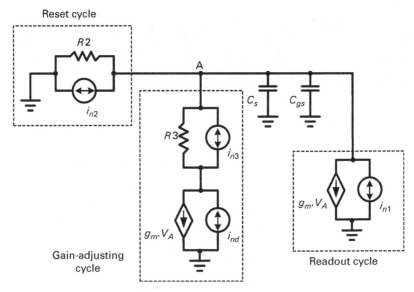

Figure 5.16 Noise model of the pixel circuit during the reset and gain-adjusting cycle.

branch (cascade branch) [94]. Thus, the noise effect of this branch is given as

$$V_{nd} = \frac{i_{nd}}{C_T s + g_{md}}, \tag{5.9}$$

in which g_{md} and i_{nd} are the trans-conductance and noise of Td, respectively. The total noise of the pixel with gain-adjusting branch is

$$V_{n-dmp} = \frac{i_{nd}}{C_T s + g_{md}} + \frac{i_{n2}}{C_T s + 1/R2} \frac{1}{\left[1 + \frac{\tau_L(V_R - V_T)}{\tau}\right]}. \tag{5.10}$$

Since A_{dmp} can be very small, the noise effect of T2 becomes negligible. Thus, by proper aspect ratio of Td, the noise performance can be almost the same as the one without the gain-adjusting branch.

(a)

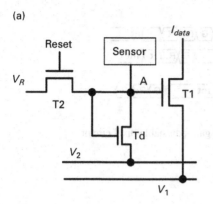

Figure 5.17 3-TFT gain-adjustable biomedical imager pixel circuit (a) and signal diagram (b).

(b)

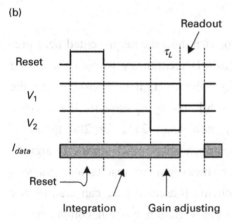

To improve the resolution, we can eliminate T3 and use Td as the storage capacitor. Figure 5.17 demonstrates the 3-TFT gain-adjustable pixel circuit. Also, the pixel provides a separate path for gain adjusting, reset, and readout; thus, the timing schedule can be improved for more parallelism as shown in Figure 5.18. Here, while the pixels in one row are being reset, the next adjacent row's pixels are tuned for the gain, and the row after that is readout. As a result, it can provide for a fast refresh rate suitable for high frame rate real-time imaging.

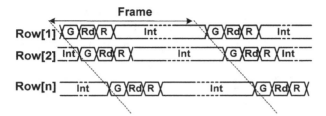

Figure 5.18 Timing schedule for 3-TFT gain-adjustable pixel circuit.

5.3 Summary

The charge-based compensation driving scheme presented here preserves the cost advantage of a-Si:H technology without increasing the implementation complexity. Moreover, it compensates for the instabilities of the backplane. While the aging error for the conventional 2-TFT pixel circuit is more than 25% for 200 hours of operation, the charge-based compensation pixel circuit is significantly more stable. Moreover, measurement results and analysis reveal that the new driving scheme presented here can control any spatial mismatch and temperature variation without compromising the settling time. The image quality of the 9-inch panel after prolonged continuous operation demonstrates the potential of this driving scheme for different applications.

Also, this technique is adapted in a biomedical imager to adjust the gain for different modalities with different signal intensities. Unlike other proposed high-dynamic-range pixel circuits [23], the readout is not destructive. Hence, it enables the use of OTRA for the entire dynamic range including high-intensity signals (e.g. radiology). Moreover, since the gain of OTRAs is independent of the readout time (unlike charge-pump amplifiers), it enables faster readout leading to higher frame rate for real-time imaging.

Copyright notices

Figures 5.2, 5.3, 5.4, 5.5, 5.6(a), and 5.11 © 2007 John Wiley & Sons. Reprinted, with permission, from [104].

Portions of the text are © 2007 John Wiley & Sons. Reprinted, with permission, from [104].

6 High-resolution architectures

For some applications such as high-resolution AMOLED displays for TVs and monitors, as well as for highly sensitive imaging modalities such as radiography and radiotherapy, the backplane should compensate for all the effects caused by the aging or mismatches to achieve the intended accuracy (e.g. less than 0.5% differential aging for a TV screen). While the entire proposed driving scheme compensates only for the static effect of V_T-shift (i.e. drop in the pixel current/gain), the transient effects such as charge injection and clock feed-through can cause up to 10% error in the pixel characteristics. To solve this issue, we introduce a calibration technique capable of controlling the transient effects as well as the static effect of the V_T-shift (mismatches) [108, 112, 113]. Moreover, a hybrid approach is introduced that takes advantage of both calibration technique and in-pixel compensation.

6.1 Time-dependent charge injection and clock feed-through

To compensate for the V_T-shift/mismatch, the compensation techniques [64, 67] lead to a modification in the gate voltage of the drive TFT to provide a constant overdrive voltage. However, as a consequence of this change in the gate voltage, clock feed-through and charge injection associated with the parasitic capacitances change over time. Since the storage capacitor cannot be large due to the mandated frame time and aperture ratio, the parasitic capacitances stemming from the overlap of the gate over the drain and source become comparable. Moreover, the threshold voltage of the switch TFTs decreases

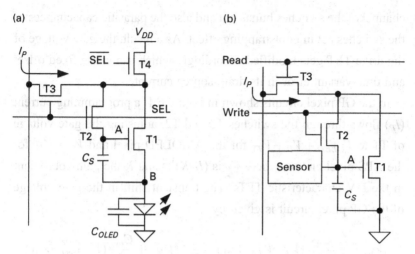

Figure 6.1 Gate-programmed pixel circuit for (a) AMOLED display [108] and (b) APS based on [109].

since they are under negative bias stress during most of the frame time [58]. Consequently, the charge profile of their channels changes for a given programming voltage inducing a time-dependent error.

The transient shifts including charge injection and clock feed-through can either increase or decrease the pixel current over time depending on the circuit topology and driving scheme. The two most probable topologies used in the design of pixel circuits, in particular for AMOLED displays, are gate-programmed (GP) and source-programmed (SP) [108]. Figure 6.1 demonstrates two GP pixel circuits for AMOLED and APS applications, respectively. In the GP pixel circuits, the source voltage is fixed and the gate voltage changes during the programming cycle to adjust the pixel current or more importantly compensate for TFT non-idealities such as non-uniformities, aging, temperature, and hysteresis. Here, while the gate voltage of the driver TFT is not transitioning after programming, the select signal of the switches changes sharply to disconnect the pixel from the common signal lines. As a result of this sharp transition, the charge in the channel of the switches bursts out and also the parasitic capacitances of the switches act in bootstrapping effect. As a result, the gate voltage of the drive TFT gets modified accordingly which can cause fixed offset and time-variant error in its drain-source current.

In the GP pixel circuits shown in Figure 6.1, a programming current (I_P) flows through the switches T3 and T2, adjusting the gate voltage of T1 to $V_{OLED} + V_P + V_{T1}$ for the AMOLED pixel and $V_P + V_{T1}$ for the APS pixel circuit, where V_P is $(I_P/K)^{1/2}$ and K the gain coefficient in the I–V characteristic TFTs. The transient shift in the gate voltage of the GP pixel circuit is given by

$$\Delta V_{gp} = -\frac{C_{g2}}{2 \cdot C_S}(V_H - V_{T2} - V_{g1}) - \frac{C_{ov2}}{C_S}(V_H) + \frac{C_{ov1}}{C_S}(V_{DD\text{-}eff} - V_{g1})$$

$$(6.1)$$

where C_{g2} is the gate capacitance of T2, C_S the storage capacitor, V_H the ON voltage of switch TFTs (OFF voltage is assumed to be zero), V_{T1} and V_{T2} the threshold voltage of T1 and T2, respectively, and V_{g1} the gate voltage of T1 which is $V_{OLED} + V_{T1} + V_P$ for Figure 6.1(a) and $V_{T1} + V_P$ for Figure 6.1(b). Here, C_{ov1} and C_{ov2} are the overlap capacitances of T1 and T2, respectively, and $V_{DD\text{-}eff}$ is the effective voltage at the drain of T1 during the driving cycle (in AMOLEDs, $V_{DD\text{-}eff} = V_{DD} - V_{DS4}$) or the readout cycle (in APSs, $V_{DD\text{-}eff}$ is a biasing voltage). The time dependence of (6.1) can be calculated as

$$\frac{d}{dt}\Delta V_{gp} = K_1 \frac{\partial}{\partial t} V_{T1} + K_2 \frac{\partial}{\partial t} V_{T2} + K_3 \frac{\partial}{\partial t} V_{OLED}$$

$$K_1 = \frac{\partial}{\partial V_{T1}}\Delta V_{gp} = \frac{1}{2 \cdot C_S}(C_{g2} - 2 \cdot C_{ov1})$$

$$K_2 = \frac{\partial}{\partial V_{T2}}\Delta V_{gp} = \frac{C_{g2}}{2 \cdot C_S}$$

$$(6.2)$$

$$K_3 = \frac{\partial}{\partial V_{OLED}}\Delta V_{gp} = \begin{cases} \frac{1}{2 \cdot C_S}(C_{g2} - 2 \cdot C_{ov1}) \\ 0 \end{cases}.$$

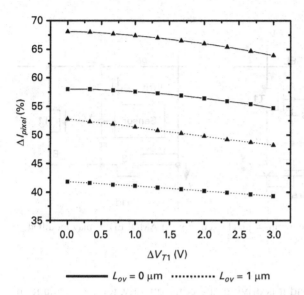

Figure 6.2 Effect of charge injection and clock feed-through on the current of the GP pixel circuit for W_{T2}=100 μm (squares) and W_{T2}=120 μm (triangles) [113].

According to (6.2), the current of the GP pixel circuits increases as it ages. Moreover, the major effect is due to C_{g2}, which is representative of the charge injection component. Figure 6.2 displays the effect of transient shifts on the current (I_{pixel}) of the GP pixel circuit in Figure 6.1(a). Cadence and a model presented by Servati [65] are used for the simulations unless a different simulator is noted. Clearly, the transient shifts include a fix offset, and a time-dependent component. Since increasing the width of T2 (W_{T2}) results in a larger parasitic capacitance, a larger T2 boosts both components of the transient effects. As indicated in (6.2) and Figure 6.2, the charge injection and clock feed-through components can compensate each other. Thus, the transient effects can be controlled significantly by increasing C_{ov1}, properly.

In the SP pixel circuits, the gate voltage is fixed and the source voltage changes during the programming cycle. In this case, after the switches turn off, the source voltage starts to adjust to its final value

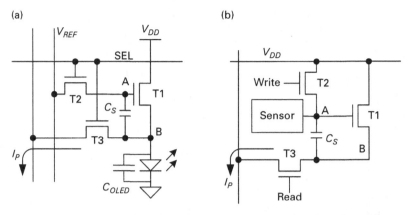

Figure 6.3 Source-programmed pixel circuit for (a) AMOLED display modified from [110] and (b) APS.

according to the load it is driving and consequently, the gate voltage of the drive TFT bootstrapped by the storage capacitor as well. Similarly to GP pixel circuits, this transition can induce fixed offset and time-dependent error into the drain-source current of the driver TFT.

In the SP pixel circuits, demonstrated in Figure 6.3, a programming current (I_P) flows from T2, and adjusts the source voltage of T1 to $V_{REF} - V_P - V_{T1}$. Here, V_{REF} is the gate voltage of T1 during the programming cycle. The transient shift in the gate voltage of the SP pixel circuit is given by [113]

$$\Delta V_{ch-sp} = -\frac{C_{g2}}{2 \cdot C_S}(V_H - V_{T2} - V_{REF}) - \frac{C_{ov2}}{C_S}(V_H + V_{OLED} + V_{T1})$$
$$-\frac{C_{ov1}}{C_S}(V_{SS-eff} + V_{T1}) \tag{6.3}$$

where C_{ov2} is the overlap capacitances of T2, and V_{SS-eff} the effective voltage at the source of T1 during the driving cycle (in AMOLEDs, $V_{SS-eff} = V_{OLED}$) or the readout cycle (in APS, V_{SS-eff} is a biasing voltage). The time dependence of (6.3) can be calculated as

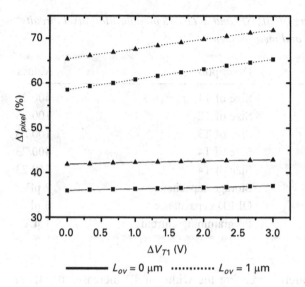

Figure 6.4 Effect of charge injection and clock feed-through on the current of the SP pixel circuit for $W_{T2}=100$ μm (squares) and $W_{T2}=120$ μm (triangles) [113].

$$\frac{d}{dt}\Delta V_{sp} = K_1 \frac{\partial}{\partial t}V_{T1} + K_2 \frac{\partial}{\partial t}V_{T2} + K_3 \frac{\partial}{\partial t}V_{OLED}$$

$$K_1 = \frac{\partial}{\partial V_{T1}}\Delta V_{sp} = -\frac{1}{C_S}(C_{ov1} + C_{ov2})$$

$$K_2 = \frac{\partial}{\partial V_{T2}}\Delta V_{sp} = \frac{C_{g2}}{2 \cdot C_S} \qquad (6.4)$$

$$K_3 = \frac{\partial}{\partial V_{OLED}}\Delta V_{sp} = \begin{cases} -\frac{1}{C_S}(C_{ov1} + C_{ov2}) \\ 0 \end{cases}.$$

According to (6.4), the current of the SP pixel circuit decreases as it ages. Unlike the GP pixel circuit, the dominant time-dependent component stems from the overlap capacitance. However, the V_{T2}-dependent part of the transient shifts is the same for both GP and SP pixel circuits. Simulation results for Figure 6.3(a) are depicted in Figure 6.4 highlighting the impact of transient shifts on the current of the SP pixel circuit. Table 6.1 lists the parameters used in the

Table 6.1 *Parameters of GP, SP, and 3-TFT step-calibration pixel circuits used in simulations and measurement.*

Name	Description	Values
$W/L(T1)$	Size of T1	400/23
$W/L(T2)$	Size of T2	100/23
$W/L(T3)$	Size of T3	100/23
$W/L(T4)$	Size of T4	400/23
$W/L(T_{OLED})$	Size of T4	750/23
C_S	Storage capacitance	2 pF
C_{OLED}	OLED capacitance	5 pF
I_P	Programming current	1 µA

simulations. Although increasing the width of T2 increases the fixed offset, its effect on the time-dependent component is negligible since T1 is much larger than T2 and so its overlap capacitance (C_{ov1}) is dominant. Moreover, as is predicted by (6.4), if the overlap capacitances are zero, the time-dependent component of the error becomes zero.

Even though most of the compensating schemes proposed to date try to compensate for the DC shift in V_T, the previous analysis shows that the transient shifts stemming from the charge injection and clock feed-through can induce significant error in the pixel current. Following the analysis, the dynamic effects can be reduced by using either a larger storage capacitor or a smaller switch TFT. However, the size of the switch TFT determines the settling time [17], and the size of the storage capacitor is limited by the aperture ratio and pixel area.

6.2 Successive calibration

The V_T-shift (ΔV_T) in the a-Si TFT under constant current stress is given by [60]

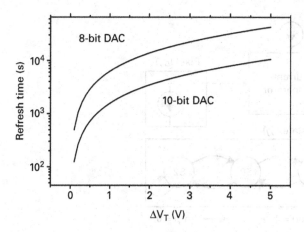

Figure 6.5 The maximum refresh time for digital calibration [112].

$$\Delta V_T = \frac{\left(\frac{I_{DS}}{K}\right)^{\frac{\gamma}{\alpha}}}{(1 + 1/\alpha)^{\gamma}} \left(\frac{t}{t_0}\right)^{\beta} \tag{6.5}$$

where, I_{DS} is the current in the TFT, K and α are the gain and power parameters in the TFT I–V characteristic, respectively, β the power law index of hydrogen escape rate, and γ the power parameter relating dangling bond creation to the band tail states [58]. Assuming that the maximum achievable accuracy of the system is set by the digital-to-analog converters (DACs) of the source driver, the refresh time of any individual pixel must be smaller than the time required for the V_T-shift to become larger than a voltage step of the driver DAC. Thus, using the linear approximation around ΔV_T, the maximum refresh time can be written as

$$t_r = \frac{V_S\left(\frac{t_\Delta}{t_0}\right)}{\beta \Delta V_T}. \tag{6.6}$$

Here, V_S is a voltage step of the driver DAC, t_Δ the corresponding time of ΔV_T, and t_r the maximum refresh time. The simulation results for two different driver DACs are shown in Figure 6.5. For γ, β, α, and t_0,

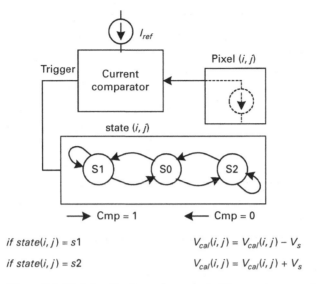

Figure 6.6 Digital calibration using a single-bit current comparator [112].

the values listed in [60] are used and the aspect ratio of the TFT is set to 400 μm/27 μm with a mobility of 0.5 cm^2/V s and $I_{DS} = 1.5$ μA. It is clear that the refresh time is larger than 100 s, and it increases as the TFT ages. Also, the non-linear part of the curves can be avoided by pre-stressing the TFT, to get a slower refresh rate. Moreover, Figure 6.5 signifies that the V_T-shift is a very slow process. Thus, by knowing the previous V_T, a single-bit ADC can follow the aging to acceptable accuracies. To do so, the pixel is programmed with a voltage as

$$V_{ref}(i,j) = V_{ref0} + V_{cal,\,n-1}(i,j) \qquad (6.7)$$

in which, $V_{ref}(i, j)$ is the reference voltage of the pixel at the ith row and jth column and $V_{cal,n-1}(i, j)$ the previous calibration voltage of pixel (i, j). Then, the current of each pixel is compared with a reference current, which can be the current of a reference pixel that is not under stress thus remaining stable. The calibration voltage is updated based on the state machine shown in Figure 6.6. Unlike the successive approximation ADC [111], the calibration voltage of each pixel is updated by only one step voltage (V_S) during each calibration turn. If Cmp is one (the estimated V_T is smaller than the real V_T), the state machine goes to $s1$. If Cmp is zero (estimated V_T is bigger than the real V_T), the state machine goes to $s4$. The small refresh time for the display ensures that the driver does not lose any aging data. For example, to calculate the refresh time, we assume that there is a comparator for each display column. During the blanking time (of the order of 500 µs) which is the free time at the end of a frame, we can extract the aging of at least 10 rows considering that the programming of each row with a reference voltage takes less than 20 µs. Hence, the refresh time for a high-definition display ($1920 \times RGB \times 1080$) with a 60 Hz frame rate is around 2 s which is much smaller than the needed 100 s.

Moreover, this technique can resolve any offset associated with the source driver since the offset has the same effect as the V_T-shift. Also, to control the offset associated with the current comparator and reference current, one can calibrate all the comparators and reference pixels with a fixed current at the beginning. For this calibration, the pixel current is replaced with a fixed current and the same algorithm shown in Figure 6.6 is used to calibrate the comparator and reference pixel circuits. Thus, V_{ref0} in (6.7) is replaced with $V_{ref0}(j)$, which is for the comparator at the jth column. To avoid increasing the source driver size, the state machine can be implemented as a firmware at the controller. To pass the results of the comparators to the controller, the shift register chain that exists in the source driver for writing the row data can be used. However, the remaining issue is the size of comparators that can result in a significant increase in the die area. Figure 6.7 displays two pixel circuits based on the step-calibration driving scheme applied to AMOLED displays and APSs. Here, the pixel current is observed through the Monitor line to extract the shift in V_T of T1.

The step-calibration proposed in Figure 6.6 benefits from simple implementation but it does not follow abrupt changes in the V_T. While

(a)

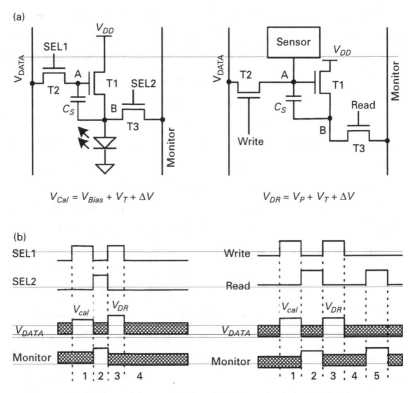

$$V_{Cal} = V_{Bias} + V_T + \Delta V \qquad\qquad V_{DR} = V_P + V_T + \Delta V$$

Figure 6.7 (a) 3-TFT AMOLED [112] and (b) AMI pixel circuits for the step-calibration driving scheme [113].

the V_T shifts gradually due to aging, it changes sharply due to temperature variations. Thus, for applications in which the temperature varies abruptly, a new driving scheme is required. A modified version of the step-calibration presented here can follow any sharp variation in V_T. To improve the algorithm, two gain stages are added to the state machine (see Figure 6.8). When the system is in state E, the previous extract V_T, $V_T(i, j)$, is applied to the pixel in the ith row and jth column. If the Trigger is zero, the system changes its state to G1, which means the actual V_T is larger than $V_T(i, j)$. At the state G1, the predicted V_T is increased by V_S. The states G2 and G3 operate similarly as G1. The only difference is that state G2 changes the operational mode to G3 if

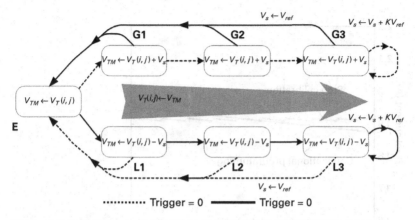

Figure 6.8 State diagram used for gained step calibration [113].

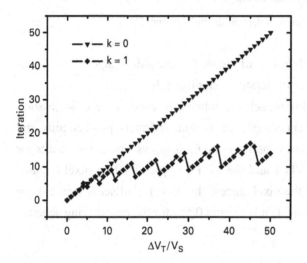

Figure 6.9 Number of iterations required for detecting ΔV_T [113].

Trigger is zero, and state G3 increases V_S intelligently to expedite the extraction of V_T-shift. States L1, L2, and L3 are the counterparts of G1, G2, and G3 for negative V_T-shift and are used when the actual V_T is smaller than the previously calculated one. Simulation results for the number of required iterations to extract the different V_T-shifts are depicted in Figure 6.9. While the number of iterations increases

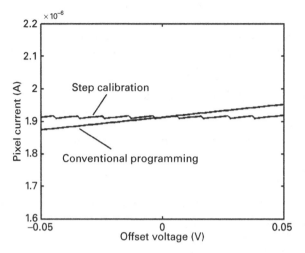

Figure 6.10 Effect of driver offset on the pixel current [113].

linearly using the algorithm of Figure 6.6, the gained algorithm reduces the required number of iterations significantly.

The proposed driving scheme inherently cancels the offset associated with the different components including drivers, pixel circuits, and up to the current comparator. Figure 6.10 shows simulation results for the conventional 2-TFT and the 3-TFT step-calibration pixel circuits. While the error in the pixel current due to 0.1 V offset is over 5% for the conventional 2-TFT, it is around 0.5% for the new driving scheme.

6.3 Arrays structure and timing

Figure 6.11(a) presents the integration method for the new pixel circuit along with the required blocks. The extraction procedure occurs in two different stages of device lifetime. First, the panel is put under calibration after fabrication and the data is stored inside the extraction memory. At this stage, timing is not an issue since the normal operation of the display is halted. The second calibration is performed during the

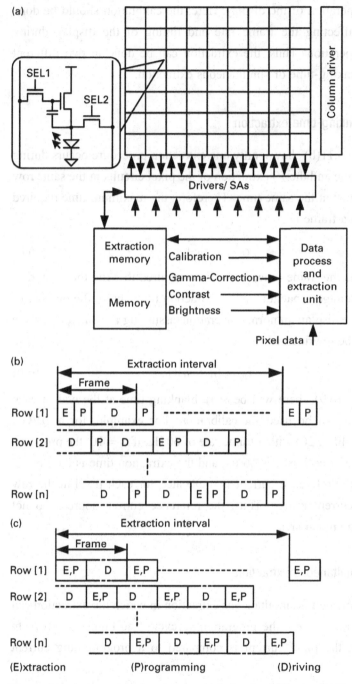

Figure 6.11 Arrays structure and proposed timing for calibration [113].

normal operation of the display. Here, the calibration should be done without affecting the frame rate and timing of the display during normal operation. Thus, the extraction can be done in two different ways: blanking-time or simultaneous extraction.

6.3.1 Blanking-time extraction

As Figure 6.11(b) shows, only one extraction procedure occurs during a frame time and the V_T extraction of the pixel circuits in the same row is performed at the same time. Therefore, the maximum time required to refresh a frame is

$$\tau_F = n \cdot \tau_P + \tau_E. \tag{6.8}$$

Here, τ_F is the frame time, τ_P the time required to write the pixel data into the storage capacitor, τ_E the extraction time, and n the number of rows in the display. In normal operation, assuming $\tau_E = m.\tau_P$, the frame time can be rewritten as

$$\tau_F = (n + m)\tau_P. \tag{6.9}$$

Following (6.9), there will be $m \cdot \tau_P$ blanking time at the end of each frame that can be used for calibration. For example, for a QVGA display (240×320) with a frame rate of 60 Hz, if $m = 10$, the programming time of each row is 66 μs, and the extraction time is 0.66 ms.

Here, the reference current is duplicated for each pixel at the row using a current mirror. Thus, the reference current sources do not occupy too much area.

6.3.2 Simultaneous extraction

Another method is simultaneous extraction in which the extraction can be performed during the programming cycle (see Figure 6.11(c)). In this case, the pixel current is compared to a programming current instead of a fixed reference current. Therefore, the frame time is $n \cdot \tau_P$, which fits in the timing budget of large-area displays easily. In this case, it is hard to duplicate the current using a current mirror since the reference current is different for each pixel. Thus, a complicated programmable current source is required at each comparator to provide the reference current. To reduce the number of programmable current sources at the source driver IC, the extraction occurs for one column or a few columns at each frame time, and so the current sources are shared between the pixels.

6.4 Configurable current comparator

Figure 6.12(a) shows a configurable current comparator [112], which operates as the output buffer and current comparator. To act as an output buffer, V_{b1} which is connected to the corresponding DAC at the source driver has the pixel luminance data, V_{b2} is V_{dd}, and CMP is zero. Therefore, the pixel is programmed through the V_{out}/I_{in} port. For comparing the current, CMP is one, and V_{out}/I_{in} port conveys the reference and pixel currents to the current comparator. To improve the comparator resolution, the input current is amplified by $(R1+R2)/R2$. Since the ratio of these resistors is important, their value can be low, thus reducing the input referred noise and die area. The signal diagram of Figure 6.12(b) is used to test the operation of the comparator.

During the reset cycle, the current of a reference pixel, which is programmed by $V_{ref0}(j)$, is conveyed into the current comparator. As a result the gate voltage of M2 is self-adjusted to allow the reference current to pass. The internal nodes of the latch (M3-M12) are also set to V_{dd}. During the sense cycle, the normal pixel circuit is connected to the current comparator through the monitor port of the pixel and the

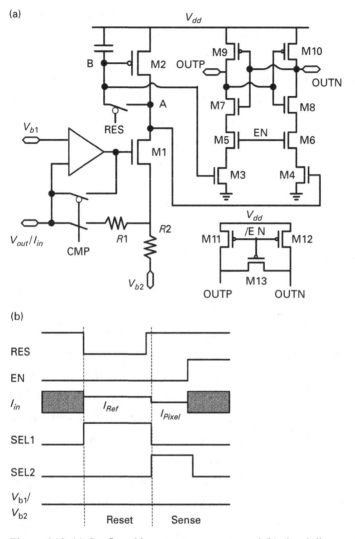

Figure 6.12 (a) Configurable current comparator and (b) signal diagram for comparator operation [112].

V_{out}/I_{in} port of the current comparator. Therefore, the change in the voltage at node A (ΔV_A) can be written as

$$\Delta V_A = \frac{R1 + R2}{R2} r_{o2}(I_{ref} - I_{pixel}). \tag{6.10}$$

Here, r_{o2} is the output resistance of M2. Then, the latch is activated and its state changes according to the difference between voltages at nodes A and B. The output of the latch can be used to adjust the calibration voltage.

During normal operation, the gate voltage of T1 is charged to $V_P + V_{T0} + \Delta V_T$ where V_P is the programming voltage and V_{T0} the initial threshold voltage of T1. Then T2 is turned off thus affecting the gate voltage of T1 by charge injection [97]. The effect of charge injection (ΔV_2) of T2 on the gate voltage of T1 can be written as

$$\Delta V_2 = -\frac{C_{gs2}}{2C_S}(V_H - V_P - V_{T0} - \Delta V_T). \qquad (6.11)$$

Here, V_H is the ON voltage of SEL1. Thus, as ΔV_T increases, the magnitude of ΔV_2 reduces. As a result the voltage remaining on the gate of T1 after turning off T2 increases, and so the pixel current increases. However, if T2 of the normal pixel circuit turns off during the sense cycle, the comparator will compare the pixel current affected by charge injection, and so this effect can be compensated by calibration. This is further explained in the next section by measurement results.

6.5 Measurement results and discussions

Figure 6.13 shows the photomicrograph of the fabricated pixel circuit and the current comparator in amorphous silicon and 0.8-μm high-voltage CMOS technologies, respectively. The diode connected transistor TO and the capacitor CO in the pixel emulate the OLED. A micro controller (PIC16F628) is used to implement the firmware and to generate the signals. Also, level shifters are used for converting the low-voltage signals to high-voltage signals. The V_{dd} and V_{ss} of the current comparator are 18 V and 0 V, respectively, while V_{dd} and V_{ss} of the pixel are 10 V and −10 V. V_{b1} and V_{b2} are set at 10 V ($V_{b1} = V_{b2}$ results in lower power consumption). For $R1$ and $R2$, 100 kΩ and

(a)

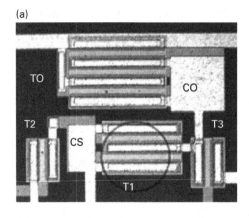

(b)

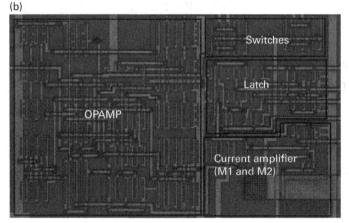

Figure 6.13 Photomicrograph of (a) fabricated pixel circuit and (b) configurable current comparator [112].

100 Ω external resistors are used, respectively. The frame rate is 60 Hz, and the programming and calibration time is 20 µs. Here a 1 × 2 array is integrated with the discrete pixels where one pixel is used as reference pixel and the other one as working pixel. The current of the working pixel is measured by using a trans-resistance amplifier connected between the ground and source of T1. Also two 100-pF external capacitors are used to emulate the parasitic capacitance of the Monitor and V_{data} lines in a real panel.

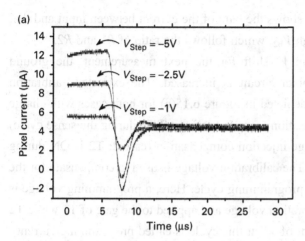

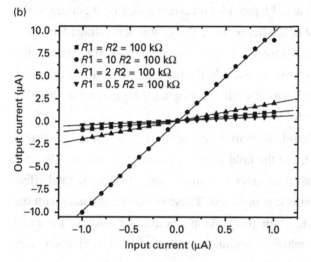

Figure 6.14 (a) Settling time for different step voltages and (b) gain of configurable current comparator [112].

Figure 6.14(a) shows the response of the pixel current to a step voltage applied to the gate of the drive TFT. As is clear, the current settles in less than 10 μs, which means that the current comparator is not a limiting factor for the refresh time. Moreover, the settling time is independent of the load at the monitor line since the monitor line is virtually grounded.

Figure 6.14(b) shows the gain of the current between input and that of passing through V_{b2}, which follows the ratio of $R1$ and $R2$.

To emulate the V_T-shift for the next measurement, the ground voltage of the pixel circuit is increased. The extracted calibration voltage (V_{cal}) is depicted in Figure 6.15(a) for both cases with charge injection compensation (case 1: turning off T2 during the sense cycle) and without charge injection compensation (case 2: T2 is ON during the sense cycle). The calibration voltage is used to compensate for the aging during the programming cycle. Here, a programming voltage is added to the calibration voltage and applied to the gate of T1 while T2 is ON (the current of T1 at this cycle is called programming current). Then T2 turns off and T1 provides a current called hold current to the OLED for the rest of the frame time. The calibration voltage for case 1 is larger than that of case 2 but the slope of case 1 is 0.8 whereas the slope for case 2 is 0.96. As a result, the programming current of case 1 starts dropping as V_T shifts while the programming current of case 2 is stable (see Figure 6.15(b)).

On the other hand, as shown in Figure 6.15(c) the hold current of case 1 is stable, but the hold current for case 2 increases. Also, the hold current of case 1 is larger than that of case 2 since the total effect of charge injection is compensated. These results corroborate with the conclusions arising from (6.7). As predicted, the error in the pixel current is 0.47% which is within the range of the quantization error of DACs at the source driver.

The successive calibration can be used to compensate for the other components of the pixel such as the OLED [114].

This technique was used in a 4.8-inch RGBW display. The results demonstrated in Figure 6.16 show compensation for backplane and OLED efficiency degradation. The display was aged with a checker board for an aggregate of four weeks at 70 °C. As can be seen the compensation works at all gray levels and different colors.

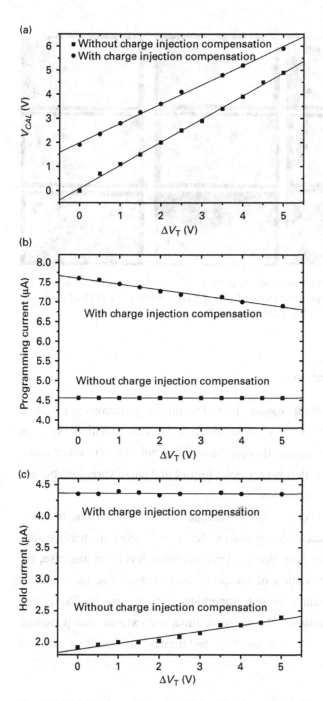

Figure 6.15 Measurement results of (a) calibration voltage (b) programming current and (c) hold current for calibration with and without charge injection compensation [112].

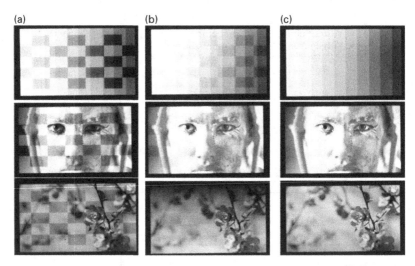

Figure 6.16 Measurement results of compensation for burnt-in checker-board pattern with different compensation levels: (a) no compensation, (b) backplane compensation and (c) backplane and OLED compensation.

6.6 Hybrid approach

While external compensation provides unique performance for large and slow variation, its complexity rises for fast variations such as hysteresis, temperature, IR drop, and cross talk. On the other hand, in-pixel compensation is very good for fast and small variations but not so good at large variations.

A good approach to take advantage of both methods is to use a pixel with a decent compensation level and boost its performance with external compensation for large variation levels. In this case, the imperfect compensation of the pixel circuits due to settling time or parasitic capacitance is not important since it can be fixed with external compensation. At the same time, the external compensation does not need fast measurement techniques for handling the fast transient effects.

6.7 Summary

The driving scheme proposed here reduces the aging effects down to 0.5% by controlling the DC and transient shifts in the V_T and gate voltage of drive/amplifier TFT. Moreover, the algorithm presented here can inherently manage the offset associated with different driving components reducing the complexity of the driver IC. Adding gain states to the extraction algorithm enables tracing of sharp variation in V_T using a single-bit comparator with a reasonable number of iterations. Taking advantage of the slow aging rate of a-Si TFTs and fast settling of voltage programming, a single-bit current-mode comparator is designed for digital calibration. The refresh time of the proposed calibration can be as low as 2 s for a high definition panel ($1920 \times 3 \times 1080$). Measurement results presented here show that the error in the pixel current after a 5-V shift in the threshold voltage of the drive TFT is less than 0.47%, which is in the range of the quantization error of the source driver.

Copyright notices

7 Summary and outlook

Despite the spatial and temporal non-uniformities associated with the TFT, implementation of stable and uniform backplanes in which the TFTs provide analog functions is described in detail with examples. The development of stable driving schemes for different applications is a critical step toward the realization of reliable and practical imagers and displays. In addition to high stability, the implementation cost, power consumption, and additive noise must be mitigated. To maximize the performance of various applications, different solutions are required since the specifications vary substantially. Thus, a set of driving schemes that can cover a wide range of intended applications for TFT backplanes is recommended.

Although the current mode active matrix has an intrinsic immunity to mismatches and differential aging, the long settling time at low current levels and large parasitic capacitance is a lingering issue, particularly for large-area applications. Consequently, a current-biased voltage-programmed (CBVP) pixel circuit that benefits from the high immunity of current programming yet has a fast settling time, low implementation cost, and low power consumption is proposed. In particular, the CBVP driving scheme is adequate for technologies that are prone to mobility as well as V_T variations. A 16×12 sensor array is fabricated with a CBVP pixel circuit that demonstrates a low noise. The array uses an operational trans-resistance amplifier (OTRA) as the readout circuitry. This enables a faster readout process and therefore real-time operation. In addition, and unlike the hybrid PPS-APS driving schemes, a gain-boosting technique based on a MIS capacitor is developed that can improve the input dynamic range from extremely low to high input signal intensities.

For very low current levels and large-area applications, a fast current driver is also developed. The settling time for a 100 nA current in the presence of a 200 pF parasitic capacitance is more than 2 ms for the conventional current source, whereas the settling time is less than 4 μs for the proposed current driver. This block can be used in a hybrid driving scheme in which current programming is used to extract the non-uniformity of the panel, while the normal operation is according to voltage programming.

For low-cost small-area AMOLED displays, a charge-based compensation technique is designed. Since on-pixel compensation is used and no change in driver requirement, the implementation cost is minimum. Measurement results of a fabricated 9-inch WXGA display show a high uniformity, despite aging the panel with a gray-scale pattern for almost 200 hours.

For high-resolution devices, a driving scheme is required that provides less than a 1% non-uniformity. As a result, all the secondary effects such as charge-injection and clock feed-through become important. Also, a successive calibration using a current comparator is proposed. This driving scheme can provide for 0.5% uniformity despite a 5-V shift in the V_T of the drive (amplifier) TFT. Also, this technique can be used in calibrating other components of the pixel such as the OLED and/or sensor.

The driving schemes presented in this book enable the use of TFT backplanes in a wide range of applications with different specifications. A quantitative comparison of these driving schemes is presented in Tables 7.1 and 7.2.

Table 7.1 *Performance comparison of different AMOLED driving schemes.*

	Compensation ability					Settling time	Power consumption	Implementation cost	Application
	V_T shift	V_T variation	Mobility variation	Temperature variation	OLED degradation				
2-TFT	– –	– –	– –	– –	–[1]	$\sim 10\ \mu s$	Low	Low	No application
VPPC	+	+	– –	– –	–[2]	$> 20\ \mu s$	Medium	Medium[3]	Small displays
CPPC	+	+	+	+	–[2]	$\sim 1\ ms$	Low	High[4]	No application
Scaling CPPC	+	+	+	+	–[2]	$> 200\ \mu s$	Medium	High[4]	No application
Additive CPPC	+	+	–	–	–[2]	$> 50\ \mu s$	High	High[4]	Small displays
CBVP	+	+	+	+	–[2]	$\sim 20\ \mu s$	Low	Low	Small-medium displays
Fast current source	+	+	+	+	–[2]	$\sim 4\ \mu s$	Low	Medium/High[5]	Large
AdMo™	+	+	+	+	–[2]	$\sim 10\ \mu s$	Low	Low	Small-medium displays
Successive calibration	+ +	+ +	+ +	+ +	+ +	$\sim 10\ \mu s$	Low	Medium[3]	Large
Relaxation	Reduced[6]	NA	NA	NA	NA	NA	NA	Low	All

Here "– –" is very poor, "–" poor, "+" good, and "+ +" very good.

[1] It does not compensate for OLED voltage shift as well as OLED luminance degradation.

[2] It compensates for only OLED voltage shift.

[3] It requires slightly modified drivers.

[4] It requires new drivers.

[5] It can be used in hybrid driving schemes in which only one current source is required to calibrate the entire display.

[6] Threshold voltage shift in TFTs is reduced and becomes saturated even under constant current operation.

Table 7.2 *Performance comparison of different biomedical driving schemes.*

	Compensation ability				Dynamic range	Input referred noise	Implementation cost
	V_T shift	V_T variation	Mobility variation	Temperature variation			
3-TFT APS	–	–	–	–	Low intensity input signals	Low	High[1]
Hybrid APS-PPS	–	–	–	–	Medium to high intensity input signals[1]	Low	Medium[2]
CBVP	+	+	+	+	Medium to high intensity input signals[1]	Lower than 3-TFTAPS	Medium[2]
Gain-boosted CBVP	+	+	+	+	Very low to high intensity input signals	Lower than 3-TFT APS	Low[3]
Gain-adjusted APS	+	+	+	+	Low to high intensity input signals	Low	Low[3]
Short-term stress	Controlled[4]	NA	NA	NA	NA	NA	NA

Here "– –" is very poor, "–" poor, "+" good, and "+ +" very good.

[1] It covers a narrow range of the applications.

[2] The cost is shared in a wider range of applications than that of 3-TFT APS.

[3] The cost is shared for most of the possible applications. Also, the driving scheme allows the use of lower cost driving circuitry such as OTRA for all range of input signal intensities.

[4] The TFTs tend to be stable under this driving scheme.

Appendix A: Enhanced voltage driving schemes

Design of the VPPCs that provides the required configurability for voltage programming is hindered by several issues: complexity (a lower yield and aperture ratio), extra controlling signals (more complex external drivers), and extra operating cycles (overhead in power consumption). Moreover, the limited time provided for V_T generation by the conventional addressing scheme, results in imperfect compensation. This appendix reviews different methods in increasing the V_T-generation time [70, 71].

A.1 Interleaved addressing scheme

The interleaved addressing scheme depicted in Figure A.1 is based on V_T generation for several rows simultaneously. The rows in a panel are divided into a few segments and the V_T-generation cycle is carried out for each segment. As a result, the time assigned to the V_T-generation cycle is extended by the number of rows in a segment leading to more precise compensation. Particularly, since the leakage current of a-Si:H TFTs is small (of the order of 10^{-14}), the generated V_T can be stored in a capacitor and be used for several other frames (see Figure A.1). As a result, the operating cycles during the following post-compensation frames are reduced to the programming and driving cycles similar to the operation of conventional 2-TFT pixel circuit [6]. Consequently, the power consumption associated with the external driver and with charging/discharging the parasitic capacitances is divided between the same few frames. In Figure A.1, the number of frames per segment is

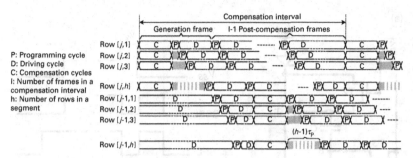

Figure A.1 Interleaving addressing scheme for low-power low-cost applications [70].

denoted as "h" and the number of frames per compensation interval as "l". As seen, the driving cycle of each row starts with a delay of τ_P from the previous row, which is the timing budget of the programming cycle. Since τ_P (of the order of 10 µs) is much smaller than the frame time (of the order 16 ms), the latency effect is negligible. However, to improve the brightness accuracy, one can either change the programming direction each time, so that the average brightness lost due to latency becomes equal for all the rows, or take into consideration this effect in the programming voltage of the frames before and after the compensation cycles.

A.1.1 3-TFT pixel circuit

Figure A.2(a) demonstrates a 3-TFT pixel circuit designed for the interleaved addressing scheme. During the first operating cycle, node A is charged to a compensating voltage. To turn off the OLED, the voltages at nodes A and C are charged to a higher voltage at the second operating cycle.

During the third operating cycle, nodes A and C are discharged through T1 until the voltage reaches V_T of T1. A programming voltage is added to the generated V_T by bootstrapping during the fourth

(a)

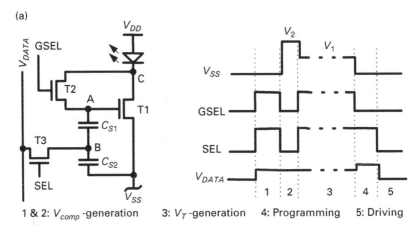

1 & 2: V_{comp} -generation   3: V_T -generation   4: Programming   5: Driving

(b)

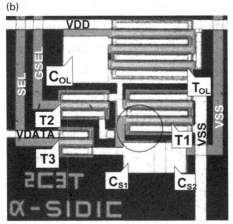

Figure A.2 (a) a-Si:H 3-TFT voltage-programmed pixel circuit, and (b) photomicrograph of the fabricated pixel circuit (TFT sizes are denoted in μm) [70].

operating cycle. Thus, the current in the pixel during the fifth operating (driving) cycle becomes independent of V_T-shift. Figure A.2(b) shows the micrograph of the fabricated 3-TFT pixel using a-Si:H technology. Here a diode connected TFT (T_{OL}) and a capacitor (C_{OL}) are used to emulate the OLED and its intrinsic capacitance respectively.

The extracted waveform for the operation of the pixel circuit using the interleaved addressing scheme is depicted in Figure A.3(a), in which the number of frames in the compensation interval is 9. The

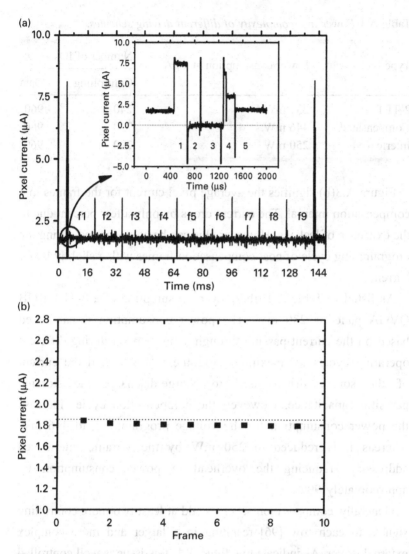

Figure A.3 (a) Measured pixel current for the interleaved addressing scheme, and (b) pixel current for different frames in the compensation interval [70].

inset plot denotes the different operating cycles including the compensation cycles. The transient cycle is explained in the next section. It is evident that the pixel current remains constant during the generation and post-compensation frames.

Table A.1 *Power and complexity of different driving schemes.*

Type	Power consumption at 1 μA	Number of IOs	
		Controlling	Data
2-TFT	230 mW	240	960
Compensated	446 mW	720	960
Interleaved	250 mW	290	960

Figure A.3(b) signifies the average pixel current for the frames in a compensation interval. The current drops by only a few percent due to the existence of leakage. However, this can be compensated during the programming cycle of post-compensation frames with a slightly larger current.

As listed in Table A.1, the power consumption of a 2-TFT RGB QVGA panel is 230 mW. The power consumption is calculated based on the current passing through a single pixel during different operating cycles for maximum brightness (1 μA) and the current of the source driver used to charge/discharge the storage/parasitic capacitance. However, the compensation cycles increase the power consumption for the voltage programming to 446 mW whereas it is reduced to 250 mW by the 9-frame interleaved addressing, reducing the overhead in power consumption by approximately 90%.

Generally, compensation schemes add at least two more controlling signals to each row [96] resulting in a larger and more complex external driver. As indicated in Table A.1, this issue is well controlled by the interleaved addressing scheme and with the array structure presented in Figure A.4. The proposed array structure diminishes the wasted area caused by extra GSEL signal by sharing VSS and GSEL signal between two physically adjacent rows. Moreover, VSS and GSEL of each row in the same segment are merged together and form

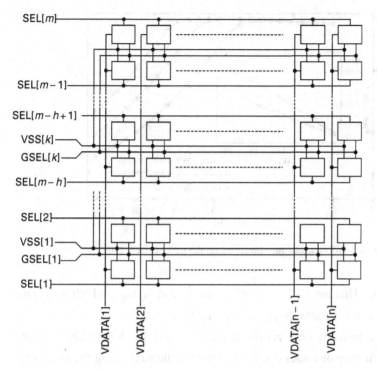

Figure A.4 Array structure sustaining the interleaved addressing scheme [70].

the segment GSEL and GVSS lines. Consequently, the controlling signals are reduced to 290 from 720 and approach 240 by including more rows in each segment. Moreover, the number of blocks driving the signals is also reduced resulting in lower power consumption and lower implementation cost.

Figure A.5 shows a set of simulation results for different values of V_2 and sizes of T2 (W2). As depicted in the inset, the larger the T2 the higher the charge injection effect and so the current rises faster due to the V_T-shift (see Chapter 6). Consequently, this can be used to compensate for OLED luminance degradation. However, a larger T2 reduces the compensating voltage developed at the gate of T1 during the third operating cycle due to higher charge injection and clock feed-through

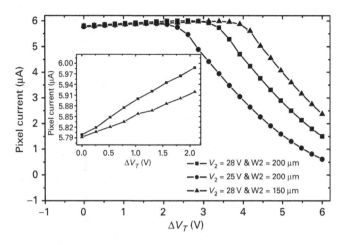

Figure A.5 Influence of non-idealities on the pixel current [70].

effects. This results in a smaller turn around voltage, which means the pixel current starts dropping for smaller V_T-shift.

The measurement results depicted in Figure A.6(a) highlight the significance of issues discussed above while supporting the simulation results shown in Figure A.5. To magnify the effects, the pixel is programmed at high current and a small voltage is used for V_2 (24 V), while V_1 is 17 V. Consequently, the circuit reaches its turn around voltage sooner. It is evident that the current increases at the beginning due to the charge injection effect. However, after 30 hours the current drops, since the pixel cannot compensate for any larger V_T-shift. However, the maximum current required by an OLED is approximately 1 µA resulting in smaller V_T-shift over time, and so the pixel continues to operate at realistic operating conditions for longer time, even with small V_2.

The measurement results in Figure A.6(b) are based on a moderated current stress level and larger V_2 (28 V). The conventional 2-TFT pixel current drops by over 70% whereas the 3-TFT pixel

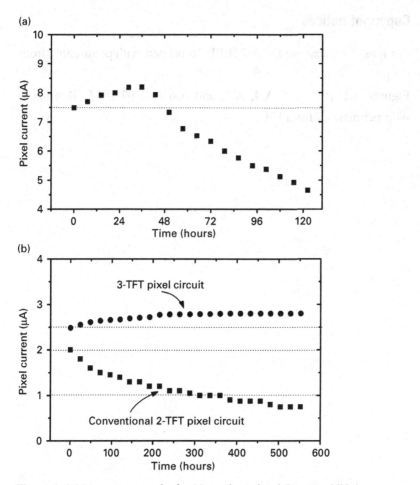

Figure A.6 Measurement results for (a) accelerated and (b) normal lifetime tests [70].

current increases by 9% which can partially compensate for OLED luminance degradation. It is expected that pixel stability can meet requirements for mobile applications ranging from cell phones to digital cameras.

Copyright notices

Appendix B: OLED electrical calibration

This appendix presents a stable compensation scheme for AMOLED displays based on the strong interdependence observed between the luminance degradation of OLED and its current drop under bias stress [114]. This feedback based compensation provides 30% improvement in the luminance stability under 1600-hour of accelerative stress. To employ this scheme in AMOLED displays, a new pixel circuit is presented that provides on-pixel electrical access to the OLED current without compromising the aperture ratio.

B.1 Interdependence between electrical and luminance degradation

Figure B.1 shows the lifetime results of a 4-mm^2 PIN-OLED with a red phosphorescent emitter for a constant luminance of 1500 nits/cm^2. A constant voltage of 3 V is applied to the OLED and the current and luminance are measured every 10 minutes. As seen, the ratio between luminance and current degradation is almost constant for the entire range. Here, ΔL and ΔI are the degradation in the OLED luminance and current, respectively, and L_0 and I_0 are the respective initial values. From the observed interdependence, it appears very likely that both luminance degradation and drop in the drive current in the OLED have a common source, and that may be the charge trapping/de-trapping at the interface of the different layers [12]. Thus, it is possible to control the aging of OLED using its output electrical

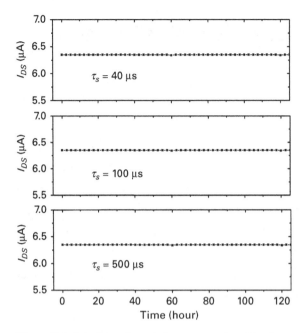

Figure B.1 Interdependence between current and luminance degradation of the OLED [114].

signals. Although, there is a deviation at degradation levels higher than 60%, we adopt a linear approximation for the compensation scheme.

B.2 Electrical compensation of OLED degradation

To investigate the effectiveness of the electrical feedback, an OLED sample is put under constant luminance stress of 1500 nits/cm². Every 30 minutes, a 3-V bias is applied to the OLED and its current is measured. Using a simple linear approximation, the OLED current stabilizing the luminance can be calculated as

$$I_P(n) = I_P(0) \left(1 + K \frac{I_B(0) - I_B(n)}{I_B(0)} \right) \tag{B.1}$$

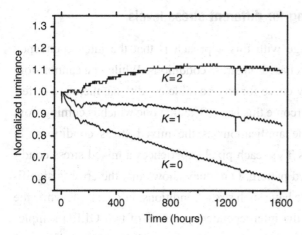

Figure B.2 Electrical feedback measurement data for OLED luminance at constant current (i.e. for $K=0$) [114].

in which K is the correction factor, $I_P(0)$ the initial programming current, $I_B(0)$ the initial biasing current following the 3-V bias, and $I_B(n)$ and $I_P(n)$ the biasing and programming currents after $n/2$ hours, respectively. Here, the current of an unstressed OLED at 3-V bias is used as $I_B(0)$ during each measurement interval. Then, the new programming current, $I_P(n)$, is applied to the OLED and its luminance is measured. Measurement results depicted in Figure B.2 show that electrical feedback compensates for the degradation by 30%. The constant current (i.e. when $K=0$) results in 40% luminance drop whereas the degradation when $K=1$ and $K=2$ is less than 10%. Here, the initial luminance is 300 nits. The correction factor and the linear approximation may vary for different OLED technologies. Nevertheless, the method shows promising improvement when the degradation in OLED current and luminance exhibit strong interdependence for a given biasing voltage.

B.3 Compensating for different stress levels

The major challenge with this approach is that the interdependency curve maybe varies based on stress conditions. While one cannot store an interdependency curve for any possible conditions, there are different methods to overcome this issue. The first one which is coming from the nature of some applications is the mixed stress conditions. In application such as TVs, each pixel experiences a mixed stress condition instead of fixed pattern. Our study shows that the error is significantly less when we have such stress conditions. For example while the variation between the interdependency curves of two OLED samples under high stress condition and low stress condition (10% brightness) could be as high as 30%, it is less than 8% when we use the mixed stress conditions (the two cases of mixed stress conditions were the following two cases. Case one: 60% of the time high, 20% of the time mid, and 20% low stress level and case two: 20% of the time high, 20% of the time mid, and 60% of the time low).

For applications that are more toward fixed patterns such as monitor, and signage, a different technique is needed. Here only a few interdependency curves based on a few fixed stress conditions are stored and each pixel value is extracted from these curves using different techniques. For example, if there are two reference curves for high and low stress (f_{high} and f_{low}), we can calculate a interdependency curve for a pixel as following:

$$St(t_i) = (St(t_{i-1}) * K_{avg} + L(t_i))/(K_{avg} + 1);$$
$$K_{high} = St(t_i)/L_{high}; K_{low} = 1 - K_{high}; \qquad \text{(B.2)}$$
$$K_{comp} = K_{high}f_{high}(\Delta I) + K_{low}f_{low}(\Delta V).$$

Here, K_{avg} is the moving average coefficient, K_{high} and K_{low} the weight of f_{high} and f_{low}, K_{comp} the compensation factor, $L(t_i)$ the luminance of the current frame, ΔV is the OLED voltage change for a fixed current (note that this parameter can be replaced with an OLED current change with a fixed voltage as well).

Also, one can change the K_{avg} based on the aging of the OLED. If OLED is aged more, that means the divergence between the two coloration curves is more. Thus, K should be larger to avoid sharp transition between the two curves. We called this technique "dynamic moving averaging."

To further improve it, one can reset the system after every aging step (or after the user turns on or off the system). We called this technique "event-based moving averaging."

$$K_{comp} = K_{comp\_evt} + K_{high}\left(f_{high}(\Delta V) - f_{high}(\Delta V_{evt})\right)$$
$$+ K_{low}\left(f_{low}(\Delta V) - f_{low}(\Delta V_{evt})\right). \tag{B.3}$$

Here, K_{comp_evt} is the compensation factor calculated at the previous event, and ΔV_{evt} is the OLED voltage change at previous event.

Figure B.3 shows a typical pixel brightness profile for more static applications. While the average brightness stays the same for a long time, there are some fluctuations around the average value due to automatic brightness control and other criteria. Also, there are major changes in the brightness due to change in the application or device operation (e.g. turning off). Figure B.3(b) shows a constructed interdependency curve based on two reference curves for the stress profile depicted in Figure B.3 using event-based moving averaging. As can be seen, while the stress condition jumps from low to high at time 9000 (arb.), the interdependency curve changes its slope to that of the high stress condition reference curve but did not jump to the curve.

Copyright notices

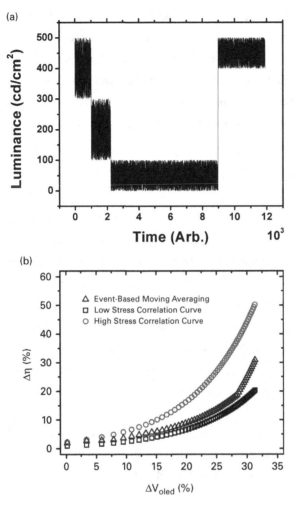

Figure B.3 (a) A typical stress condition of a pixel and (b) reconstruction of its interdependency curves based on two reference curves.

References

[1] L. E Antonuk, J. Boudry, J. Yorkston, *et al.*, "Development of thin-film flat-panel arrays for diagnostic and radiotherapy imaging," *Proc. of SPIE*, vol. 1651, 1992, pp. 94–105.

[2] W. Zhao, I. Blevis, S. Germann, and J. Rowlands, "Digital radiology using active matrix readout amorphous selenium: construction and evaluation of a prototype real-time detector," *J. Med. Phys.*, vol. 24, no. 12, pp. 1834–1843, Dec. 1997.

[3] N. Matsuura, W. Zhao, Z. Huang, and J. Rowlands, "Digital radiology using active matrix readout: amplified pixel detector array for fluoroscopy," *J. Med. Phys.*, vol. 26, no. 5, pp. 672–681, May 1999.

[4] R. M. Dawson and M.G. Kane, "Pursuit of active matrix organic light emitting diode displays," *Dig. Tech. Papers, SID Int. Symp.*, San Jose, June 5–7 2001, pp. 372–375.

[5] G. Gu and S. R. Forest, "Design of flat-panel displays based on organic light-emitting devices," *IEEE J. Sel. Topics in Quantum Elecs.*, vol. 4, pp. 83–99, Jan. 1998.

[6] A. Nathan, G. R. Chaji, and S. J. Ashtiani, "Driving schemes for a-Si and LTPS AMOLED displays," *IEEE J. Display Tech.*, vol. 1, pp. 267–277, Dec. 2005.

[7] E. Lueder, *Liquid Crystal Displays*, John Wiley & Sons, 2001.

[8] G. He, M. Pfeiffer, and K. Leo, "High-efficiency and low-voltage p-i-n electrophosphorescent organic light-emitting diodes with double-emission layers," *Appl. Phys. Letts.*, vol. 85, no. 17, pp. 3911–3913, Oct. 2004.

[9] Y. Yang and J. Bharathan, "Ink-jet printing technology and its application in polymer multicolor EL displays," *Dig. Tech. Papers, SID Int. Symp.*, Anaheim, May 1998, pp. 19–22.

[10] Z. D. Popovic and H. Aziz, "Reliability and degradation of small molecule-based organic light-emitting devices (OLEDs)," *IEEE J. on Sele. Topics in Quantum Elecs.*, vol. 8, no. 2, pp. 362–371, Mar. 2002.

[11] I. D. Parker, Y. Cao, and C. Y. Yang, "Lifetime and degradation effects in polymer light-emitting diodes," *J. Appl. Phys.*, vol. 85, no. 4, pp. 2441–2447, Feb. 1999.

[12] H. Aziz, Z. D. Popovic, N. Hu, A. Hor, and G. Xu, "Degradation mechanism of small molecule-based organic light-emitting devices," *Science*, vol. 283, pp. 1900–1902, Mar. 1999.

[13] D. Zoua and T. Tsutsui, "Voltage shift phenomena introduced by reverse-bias application in multilayer organic light emitting diodes," *J. Appl. Phys.*, vol. 87, no. 4, pp. 1951–1956, Feb. 2000.

[14] A. Nathan, A. Kumar, K. Sakariya, *et al.*, "Amorphous silicon thin film transistor circuit integration for organic LED displays on glass and plastic," *IEEE J. Solid State Cirs.*, vol. 39, pp. 1477–1486, 2004.

[15] J. K. Mahon, "History and status of organic light-emitting device (OLED) technology for vehicular applications," *Dig. Tech. Papers, SID Int. Symp.*, San Jose, June 2001, pp. 22–25.

[16] S. E. Lee, W. S. Oh, S. C. Lee, and J. C. Chol, "Development of a novel current controlled organic light emitting diode (OLED) display driver IC," *IEICE Tran.*, vol. E85, no. 11, pp. 1940–1944, Dec. 2002.

[17] S. J. Ashtiani, Pixel circuits and driving schemes for active-matrix organic light-emitting diode displays, Ph.D. Thesis, University of Waterloo, 2007.

[18] M. Baldo, The electronic and optoelectronic properties of amorphous organic semiconductors, Ph.D. Thesis, Princeton University, 2001.

[19] C. Qiu, H. Peng, H. Chen, *et al.*, "Top-emitting organic light-emitting diode using nanometer platinum layers as hole injector," *Dig. Tech. Papers, SID Int. Symp.*, Baltimore, 2003, pp. 974–977.

[20] C. Lee, D. Moon, and J. Han, "Top emission organic light emitting diode with Ni anode," *Dig. Tech. Papers, SID Int. Symp.*, Baltimore, 2003, pp. 533–535.

[21] J. Chang, Sensor system for high throuput fluorescent bio-assays, Ph.D. Thesis, University of Waterloo, 2007.

[22] K. Karim, Pixel architectures for digital imaging using amorphous silicon technology, Ph.D. Thesis, University of Waterloo, 2007.

[23] M. H. Izadi and K. S. Karim, "High dynamic range pixel architecture for advanced diagnostic medical imaging applications," *J. Vac. Sci. Tech. A*, vol. 24, no. 3, pp. 846–849, Feb. 2007.

[24] F. Taghibakhsh and K. S. Karim, "High dynamic range 2-TFT amplifier pixel sensor architecture for digital mammography tomosynthesis," *IET Cirs. Devs. Syst.*, vol. 1, pp. 87–92, Feb. 2007.

[25] K. S. Karim, A. Nathan, M. Hack, and W. I. Milne, "Drain-bias dependence of threshold voltage stability of amorphous silicon TFTs," *IEEE Elec. Dev. Letts.*, vol. 25, no. 4, pp. 188–190, Apr. 2004.

[26] A. Nathan, D. Striakhilev, R. Chaji, *et al.*, "Backplane requirements for active matrix organic light emitting diode displays," *Proceedings of MRS 2006*, San Francisco, US, Apr. 2006, pp. 0910-A16–01-L09–01.

[27] A. G. Lewis, D. D. Lee, and R. H. Bruce, "Polysilicon TFT circuit design and performance," *IEEE J. Solid State Cirs.*, vol. 27, pp. 1833–1842, Dec. 1992.

[28] M. Stewart, R. Howell, L. Dires, and K. Hatalis, "Polysilicon TFT technology for active matrix OLED displays," *IEEE Trans. on Elec. Devs.*, vol. 48, pp. 845–851, 2001.

[29] M. J. Yang, C. H. Chien, Y. H. Lu, *et al.*, "High-performance and low-temperature-compatible p-channel polycrystalline-silicon TFTs using hafnium-silicate gate dielectric," *IEEE Elec. Dev. Letts.*, vol. 28, pp. 902–904, 2007.

[30] P. Servati, Amorphous silicon TFTs for mechanically flexible electronics, Ph.D. Thesis, University of Waterloo, 2004.

[31] H. Watanabe, "Statistics of grain boundaries in polysilicon," *IEEE Trans. on Elec. Devs.*, vol. 54, pp. 38–44, Jan. 2007.

[32] Y. H. Tai, S. C. Huang, W. P. Chen, *et al.*, "A statistical model for simulating the effect of LTPS TFT device variation for SOP applications," *J. Display Tech.*, vol. 3, pp. 426–434, Dec. 2007.

[33] R. A. Street, *Hydrogenated Amorphous Silicon*, Cambridge University Press, 1991.

[34] W. B. Jackson, M. Marshall, and M. D. Moyer, "Role of hydrogen in the formation of metastable defects in hydrogenated amorphous silicon," *Phys. Rev. B*, vol. 39, no. 2, pp. 1164–1179, Jan. 1989.

[35] Y.-H. Tai, J.-W. Tsai, H.-C. Cheng, and F.-C. Su, "Instability mechanisms for the hydrogenated amorphous silicon thin-film transistors with negative and positive bias stresses on the gate electrodes," *Appl. Phys. Letts.*, vol. 67, pp. 76–78, July 1995.

[36] I. C. Cheng and S. Wagner, "High hole and electron field effect mobilities in nanocrystalline silicon deposited at 150 °C," *Elsevier Thin Solid Films*, vol. 427, pp. 56–59, Jan. 2003.

[37] M. R. Esmaeili-Rad, A. Sazanov, and A. Nathan, "Absence of defect state creation in nanocrystalline silicon thin film transistors deduced from constant current stress measurements," *Appl. Phys. Letts.*, vol. 91, pp. 113511 (1–3), Sept. 2007.

[38] M. R. Esmaeili-Rad, F. Li, A. Sazanov, and A. Nathan, "Stability of nanocrystalline silicon bottom-gate thin film transistors with silicon nitride gate dielectric," *J. Appl. Phys.*, vol. 102, pp. 064512 (1–7), Sept. 2007.

[39] Y. Y. Lin, D. J. Gundlach, S. F. Nelson, and T. N. Jakson, "Stacked pentacene layer organic thin-film transistors with improved characteristics," *IEEE Elec. Dev. Letts.*, vol. 18, pp. 606–608, Dec. 1997.

[40] F. M. Li, A. Nathan, Y. Wu, and B. S. Ong, "Organic thin-film transistor integration using silicon nitride gate dielectric," *Appl. Phys. Letts.*, vol. 90, pp. 133514 (1–3), Mar. 2007.

[41] J. F. Wager, "Transparent electronics," *Science*, vol. 300, pp. 1245–1246, 2003.

[42] K. Nomura, H. Ohta, A. Takagi, *et al.*, "Room-temperature fabrication of transparent flexible thin-film transistors using amorphous oxide semiconductors," *Nature*, vol. 432, pp. 488–492, 2004.

[43] R. Martins, A. Nathan, R. Barros, *et al.*, "Complementary metal oxide semiconductor technology with and on paper, *Adv. Mater.*, vol. 23, pp. 4491–4496, 2011.

[44] A. Nathan, S. Lee, S. Jeon, I. Song, U-In Chung, "Amorphous oxide TFTs: Progress and issues," *SID Symp. Dig. Tech. Papers*, vol. 43, no. 1, pp. 1–4, June 2012.

[45] K. Ghaffarzadeh, A. Nathan, J. Robertson, *et al.*, "Persistent photoconductivity in Hf-In-Zn-O thin film transistors," *Appl. Phys. Letts.*, vol. 97, 143510 (1–3), 2010.

[46] M. D. H. Chowdhury, P. Migliorato, and J. Jang, "Light induced instabilities in amorphous indium–gallium–zinc–oxide

thin-film transistors," *Appl. Phys. Lett.*, vol. 97, pp. 173506, 2010.

[47] S. Jeon, S.-E. Ahn, I. Song, *et al.*, "Gated three-terminal device architecture to eliminate persistent photoconductivity in oxide semiconductor photosensor arrays," *Nature Mater.*, DOI: 10.1038/ NMAT3256 (Feb. 2012), pp. 1–5.

[48] S. Lee, K. Ghaffarzadeh, A. Nathan, *et al.*, "Trap-limited and percolation conduction mechanisms in amorphous oxide semiconductor thin film transistors," *Appl. Phys. Letts.*, vol. 98, pp. 203508, 2011.

[49] M. Mativenga, M. H. Choi, J. W. Choi, and J. Jang, "Transparent flexible circuits based on amorphous-indium–gallium–zinc–oxide thin-film transistors," *IEEE Electron Device Letts.*, vol. 32, p. 170, 2011.

[50] R. Martins, A. Ahnood, N. Correia, *et al.*, "Recyclable, flexible, low power oxide electronics," *Adv. Funct. Mater.*, DOI: 10.1002/adfm. 201202907.

[51] J.-S. Park, T.-W. Kim, D. Stryakhilev, *et al.*, "Flexible full color organic light-emitting diode display on polyimide plastic substrate driven by amorphous indium gallium zinc oxide thin-film transistors," *Appl. Phys. Letts.*, vol. 95, pp. 013503, 2009.

[52] S. I. Kim, S. W. Kim, J. C. Park, *et al.*, "Highly sensitive and reliable X-ray detector with HgI2 photoconductor and oxide drive TFT," *Tech. Dig., IEEE Electron Devices Meeting (IEDM)*, 2011, DOI: 10.1109/IEDM.2011.6131550, pp. 14.2.1–14.2.4.

[53] S. Jeon, S. E. Ahn, I. Song, *et al.*, "Dual gate photo-thin film transistor with high photoconductive gain for high reliability, and low noise flat panel transparent imager," *Tech. Dig., IEEE Electron Devices Meeting (IEDM)*, 2011, DOI: 10.1109/IEDM.2011.6131551, pp. 14.3.1–14.3.4.

[54] H. W. Zan and S. C. Kao, "The effect of drain-bias on the threshold voltage instability in organic TFTs," *IEEE Elec. Dev. Letts.*, vol. 92, pp. 155–157, Feb. 2008.

[55] T. H. Kim, C. G. Han, and C. K. Song, "Instability of threshold voltage under constant bias stress in pentacene thin film transistors employing polyvinylphenol gate dielectric," *Elsevier Thin Solid Films*, vol. 516, pp. 1323–1326, June 2007.

[56] G. Gu and M. G. Kane, "Moisture induced electron traps and hysteresis in pentacene-based organic thin-film transistors," *Appl. Phys. Letts.*, vol. 92, pp. 053305 (1–3), Feb. 2008.

[57] W. F. Aerts, S. Verlaak, and P. Heremans, "Design of an organic pixel addressing circuit for an active-matrix OLED display," *IEEE Elec. Dev. Letts.*, vol. 49, pp. 2124–2126, Dec. 2002.

[58] M. J. Powell, C. Berkel, and J. R. Hughes, "Time and temperature dependence of instability mechanisms in amorphous silicon thin-film transistors," *J. Appl. Phys.*, vol. 54, pp. 1323–1325, Jan. 1989.

[59] F. R. Libsch and J. Kanicki, "Bias-stress-induced stretched-exponential time dependence of charge injection and trapping in amorphous thin-film transistors," *Appl. Phys. Letts.*, vol. 62, pp. 1286–1288, Mar. 1993.

[60] S. M. Jahinuzzaman, A. Sultana, K. Sakariya, P. Servati, and A. Nathan, "Threshold voltage instability of amorphous silicon thin-film transistors under constant current stress," *Appl. Phys. Letts.*, vol. 87, pp. 023502 (1–3), July 2005.

[61] C. S. Chiang, J. Kanicki, and K. Takechi, "Electrical instability of hydrogenated amorphous silicon thin-film transistors for active-matrix liquid-crystal displays," *Jpn. J. Appl. Phys.*, vol. 37, pp. 4704–4710, Sept. 1998.

[62] S. Sambandan, L. Zhu, D. Striakhilev, P. Servati, and A. Nathan, "Markov model for threshold-voltage shift in amorphous silicon TFTs for variable gate bias," *IEEE Elec. Dev. Letts.*, vol. 26, pp. 375–377, June 2005.

[63] Y. He, R. Hattori, and J. Kanicki, "Improved a-Si:H TFT circuits for active-matrix organic light emitting displays," *IEEE Trans. Elect. Devs.*, vol. 48, no. 7, pp. 1322–1325, July 2001.

[64] N. Safavian, G. R. Chaji, S. J. Ashtiani, A. Nathan, and J. A. Rowlands, "Self-compensated a-Si:H detector with current-mode readout circuit for digital x-ray fluoroscopy," *Proc. of IEEE MIDWEST*, Cincinnati, USA, Aug. 2005.

[65] P. Servati and A. Nathan, "Modeling of the static and dynamic behavior of hydrogenated amorphous silicon thin-film transistors," *J. Vac. Sci. Tech.*, vol. 20, no. 3, pp. 1038–1042, May 2002.

[66] J. H. Baek, M. Lee, J. H. Lee, *et al.*, "A current-mode display driver IC using sample-and-hold scheme for QVGA full-color active matrix

organic LED mobile displays," *IEEE J. Solid State Cirs.*, vol. 41, no. 12, pp. 2974–2982, Dec. 2006.

[67] Y. C. Lin, H. P. Shieh, and J. Kanicki, "A novel current-scaling a-Si:H TFTs pixel electrode circuit for AM-OLEDs," *IEEE Trans. Elect. Devs.*, vol. 52, pp. 1123–1132, June 2005.

[68] S. Ono and Y. Kobayashi, "An accelerative current-programming method for AM-OLED," *IEICE Trans. Elecs.*, vol. E88-C, pp. 264–269, Feb. 2005.

[69] G. R. Chaji, S. Ashtiani, S. Alexander, *et al.*, "Pixel circuits and drive schemes for large-area a-Si AMOLED," *IDMC* 2005, Taiwan, 2005.

[70] G. R. Chaji and A. Nathan, "Low-power low-cost voltage-programmed a-Si:H AMOLED display for portable devices," *IEEE J. Display Tech.*, vol. 4, no. 2, pp. 233–237, June 2008.

[71] G. R. Chaji and A. Nathan, "Parallel addressing scheme for voltage-programmed active matrix OLED displays," *IEEE Trans. on Elec. Devs.*, vol. 54, pp. 1095–1100, May 2007.

[72] J. L. Sanford and F. R. Libsch, "TFT AMOLED pixel circuits and driving methods," *Dig. Tech. Papers, SID Int. Symp.*, Baltimore, 2003, pp. 10–13.

[73] G. R. Chaji, P. Servati, and A. Nathan, "Driving scheme for stable operation of 2-TFT a-Si AMOLED pixel," *IEE Electronics Letts.*, vol. 41, no. 8, pp. 499–500, Apr. 2005.

[74] J. C. Goh, J. Jang, K. S. Cho, and C. K. Kim, "A new a-Si:H thin-film transistor pixel circuit for active-matrix organic light-emitting diodes," *IEEE Elect. Dev. Letts.*, vol. 24, pp. 583–585, Sept. 2003.

[75] J. C. Goh, C. K. Kim, and J. Jang, "A novel pixel circuit for active-matrix organic light-emitting diodes," *Dig. Tech. Paper, SID Int. Symp.*, Baltimore, 2003, pp. 494–497.

[76] S. W. Tam, Y. Matsueda, M. Kimura, *et al.*, "Poly-Si driving circuits for organic EL displays," *Proc. of SPIE*, vol. 4295, Apr. 2001, pp. 125–133.

[77] J. C. Goh, J. Jang, K. S. Cho, and C. K. Kim, "A new pixel circuit for active matrix organic light emitting diodes," *IEEE Elec. Dev. Letts.*, vol. 23, pp. 583–585, Sept. 2002.

[78] S. H. Jung, W. J. Nam, and M. K Han, "A new voltage-modulated AMOLED pixel design compensating for threshold voltage variation in poly-Si TFTs," *IEEE Elect. Dev. Letts.*, vol. 25, pp. 690–692, Oct. 2004.

[79] R. M. A. Dawson, *et al.*, "A polysilicon active matrix organic light emitting diode display with integrated drivers," *Dig. Tech. Papers, SID Int. Symp.*, 1999.

[80] G. R. Chaji and A. Nathan, "Stable voltage-programmed pixel circuit for AMOLED displays," *IEEE J. Display Tech.*, vol. 2, pp. 347–358, Dec. 2006.

[81] D. A. Fish, *et al.*, "Improved optical feedback for OLED differential ageing correction," *J. SID*, vol. 13, pp. 131–138, 2005.

[82] S. J. Ashtiani and A. Nathan, "A driving scheme for active-matrix organic light-emitting diode displays based on feedback," *IEEE J. Display Tech.*, vol. 2, pp. 258–264, Sept. 2006.

[83] S. J. Ashtiani and A. Nathan, "A driving scheme for AMOLED displays based on current feedback," *Proc. of IEEE CICC*, Sept. 2006, pp. 289–292.

[84] K. Inukai, *et al.*, "4.0-in. TFT-OLED displays and a novel digital driving scheme," *Dig. Tech. Papers, SID Int. Symp.*, 2000, pp. 924–927.

[85] D. Y. Kondakov, W. C. Lenhart, and W. F. Nichols, "Operational degradation of organic light-emitting diodes: mechanism and identification of chemical products," *J. Appl. Phys.*, vol. 101, pp. 024512 (1–7), Jan. 2007.

[86] G. R. Chaji and A. Nathan, "A novel driving scheme for high-resolution large-area a-Si:H AMOLED displays," *Proc. of IEEE MIDWEST*, Cincinnati, Aug. 2005, pp. 782–785.

[87] G. R. Chaji and A. Nathan, "A sub-µA fast-settling current programmed pixel circuit for AMOLED displays," *IEEE European Solid State Cirs. (ESSCIRC 07)*, Sept. 2007, pp. 344–347.

[88] G. R. Chaji, D. Striakhilev, and A. Nathan, "A novel a-Si:H AMOLED pixel circuit based on short-term stress stability of a-Si:H TFTs," *IEEE Elec. Dev. Letts.*, vol. 26, pp. 737–739, Oct. 2005.

[89] G. R. Chaji, N. Safavian, and A. Nathan, "Stable a-Si:H circuits based on short-term stability of amorphous silicon TFTs," *J. Vac. Sci. and Tech. A*, vol. 24, pp. 875–878, May 2006.

[90] I. Bloom and Y. Nemirovsky, "1/f noise reduction of metal-oxide-semiconductor transistors by cycling from inversion to accumulation," *Appl. Phys. Letts.*, vol. 58, pp. 1664–1666, Apr. 1991.

[91] E. A. M. Klumperink, J. Gerkink, A. P. van der Wel, and B. Nauta, "Reducing MOSFET 1/f noise and power consumption by switched biasing," *IEEE J. Solid State Cirs.*, vol. 35, pp. 994–1001, July 2000.

[92] A. Hassibi and T. H. Lee, "A programmable electrochemical biosensor array in 0.18µm standard CMOS," *ISSCC Dig. Tech. Papers*, pp. 564–566, Feb. 2005.

[93] E. A. H. Hall, *Biosensors*, Open University Press, 2003.

[94] M. H. Izadi and K. S. Karim, "Noise optimization of an active pixel sensor for real-time digital x-ray fluoroscopy," *Proc. of the SPIE on Noise and Flucs. in Cir., Devices, and Materials*, vol. 6600, pp. 66000Y, 2007.

[95] G. R. Chaji and A. Nathan, "A sub-µA fast-settling current programmed pixel circuit for AMOLED displays," *IEEE European Solid State Cirs. (ESSCIRC 07)*, Sept. 2007, pp. 344–347.

[96] J. H. Jung, et al., "A 14.1 inch full color AMOLED display with top emission structure and a-Si backplane," *Dig. of Tech. Papers, SID Int. Symp.*, 2005, pp. 1538–1541.

[97] G. Wegmann, E. A. Vitoz, and F. Rahali, "Charge injection in analog MOS switches," *IEEE J. Solid State Cirs.*, vol. Sc-22, pp. 1091–1097, Dec. 1987.

[98] G. R. Chaji and A. Nathan, "High-precision, fast current source for large-area current-programmed a-Si flat panels," *Proc. of IEEE ISCASS*, June 2006, Greece, pp. 541–544.

[99] G. R. Chaji and A. Nathan, "Fast and offset-leakage insensitive current mode line driver for active matrix displays and sensors," *IEEE J. Display Tech.*, vol. 5, no. 2, pp. 72–79, Feb. 2009.

[100] C. Toumazou, F. J. Lidgey, and D. G. Haigh, *Analogue IC Design: the Current-Mode Approach*, Peter Peregrinus Ltd., 1990, pp. 93–126.

[101] A. S. Sedra, G. W. Roberts, and F. Gohh, "The current conveyor: history, progress and new results," *IEE Proc. G., Elec. Cirs. Syst.*, vol. 137, no. 2, pp. 78–87, 1990.

[102] J. E. Slotine and W. Li, *Applied Nonlinear Control*, Prentice-Hall, 1991, pp. 40–99.

[103] G. R. Chaji and A. Nathan, "Low-cost stable a-Si:H AMOLED display for portable applications," *Proc. of IEEE NEWCAS*, Ottawa, Canada, June 2006, pp. 97–100.

[104] G. R. Chaji, S. Alexander, A. Nathan, C. Church, and S. J. Tang, "A low-cost amorphous silicon AMOLED display with full V_T- and V_{OLED}-shift compensation," *Tech. Dig. SID Symp.*, Long Beach, US, May 2007, pp. 1580–1583.

[105] C. Ng and A. Nathan, "Temperature characterization of a-Si:H thin-film transistor for analog circuit design using hardware description language modeling," *J. Vac. Sci. and Tech. A*, vol. 24, pp. 883–887, May 2006.

[106] C. Poynton, *Digital Video and HDTV Algorithms and Interfaces*, Morgan Kaufmann Publishers, 2007.

[107] B. Razavi, *Design of Analog CMOS Integrated Circuits*, McGraw Hill Higher Education, 2001.

[108] G. R. Chaji, N. Safavian, and A. Nathan, "Dynamic-effect compensating technique for stable a-Si:H AMOLED displays," *Proc. of IEEE MIDWEST*, Cincinnati, USA, Aug. 2005, pp. 786–789.

[109] N. Safavian, G. R. Chaji, A. Nathan, and J. A. Rowlands, "Three-TFT image sensor for real-time digital X-ray imaging," *IEE Elec. Letts.*, vol. 42, no. 3, pp. 31–32, Feb. 2006.

[110] S. J. Ashtiani, P. Servati, D. Striakhilev, and A. Nathan, "A 3-TFT current-programmed pixel circuit for active-matrix organic light-emitting diode displays," *IEEE Trans. Elect. Devs.*, vol. 52, pp. 1514–1518, July 2005.

[111] D. Johns and K. Martin, *Analog Integrated Circuit Design*, New York, John Wiley & Sons, 1997, pp. 487–530.

[112] G. R. Chaji and A. Nathan, "A current-mode comparator for digital calibration of amorphous silicon AMOLED displays," *IEEE Trans. on Cirs. and Sys. II*, vol. 55, no. 7, pp. 614–618, July 2008.

[113] G. R Chaji, N. Safavian, and A. Nathan, "Dynamic effect compensating technique for DC and transient instability of thin film transistor circuits for large-area devices," *Springer Analog Int. Cir. and Sig. Proc.*, vol. 56, no. 1–2, pp. 143–151, Aug. 2008.

[114] G. R. Chaji, C. Ng, A. Nathan, *et al.*, "Electrical compensation of OLED luminance degradation," *IEEE Elect. Dev. Letts.*, vol. 28, pp. 1108–1110, Dec. 2007.

Index

Figures and tables are referenced with bold numbers.

Printed in the United States
by Baker & Taylor Publisher Services